Understanding and Negotiating EPC Contracts

Volume 1

In *Understanding and Negotiating EPC Contracts, Volume 1*, Howard M. Steinberg presents a practical and comprehensive guide to understanding virtually every aspect of engineering, procurement and construction (EPC) contracts for infrastructure projects.

The 25 chapters in Volume 1 are supplemented with real-life examples and court decisions, and offer tactical advice for anyone who must negotiate or understand EPC contracts in connection with the implementation, financing or operation of infrastructure projects.

Emphasizing current market practices and strategic options for risk sharing, the book contains a narrative explanation of the underpinning of all of the issues involved in EPC contracting. Exhaustive in scope, it clarifies the fundamental commercial principles and pitfalls of "turnkey" contracting for all types of capital investments ranging from electrical and thermal power generation (including combined heat and power, nuclear, wind, solar, natural gas and coal) to refining, to chemical processing, LNG liquefaction and re-gasification, to high speed rail, bridging, tunneling and road building. Providing clear and thorough analyses of the issues and challenges, this volume will be of great value to all those involved in complex construction projects.

Howard M. Steinberg is of counsel and a retired partner in the law firm Shearman & Sterling LLP and has more than 25 years of legal experience in the infrastructure sector in more than 100 countries. He is named a leading lawyer in project finance by Chambers & Partners and IFLR 1000. He represents sponsors, offtakers, miners, suppliers, engineering firms, consultants, financial advisors, contractors, operators, underwriters, lenders, export credit agencies and multi-lateral institutions in the development, acquisition, restructuring and privatization of projects around the world. He holds bachelor's and business degrees from Columbia University and a law degree from New York University.

Understanding and Negotiating EPC Contracts

Volume 1

The Project Sponsor's Perspective

Howard M. Steinberg

Routledge
Taylor & Francis Group

LONDON AND NEW YORK

First published in paperback 2024

First published 2017
by Routledge
4 Park Square, Milton Park, Abingdon, Oxon OX14 4RN

and by Routledge
605 Third Avenue, New York, NY 10158

Routledge is an imprint of the Taylor & Francis Group, an informa business

British Library Cataloguing in Publication Data
A catalogue record for this book is available from the British Library

Library of Congress Cataloguing in Publication Data
Steinberg, Howard M. author.
Understanding and negotiating EPC contracts / by Howard M. Steinberg.
   pages   cm
Includes bibliographical references and index.
ISBN 978-1-4724-1106-8 (hardback)   1. Construction contracts–United States.
2. Construction industry–Law and legislation–United States.   I. Title.
KF902.S74 2014
343.7307'862131–dc23

ISBN: 978-1-4724-1106-8 (hbk)
ISBN: 978-1-03-283698-0 (pbk)
ISBN: 978-1-315-54930-9 (ebk)

DOI: 10.4324/9781315549309

Typeset in Baskerville
by Out of House Publishing

Contents

Volume I

Volume 2

Figures

Reader's Note

Thomas Alva Edison dreamed of harnessing our sun's power. Today, 60-story solar "power towers" inject millions of megawatts into electricity grids from California to Spain.

But, as many have learned, even this seemingly boundless energy source and its support from governmental funding, tax incentives, and free-flowing market capital is still vulnerable to simple budgetary misprojections and poor project management. As project and corporate failures around the world mount in every aspect of the solar industry, only the organized projects will survive.

For me, this book is about only one thing—being organized. Nothing can defend against project failure like effective organization. From my beginnings as a young "dirt" lawyer handling water rights, I have always been grateful that I have been able to watch the birth of the commercial solar industry and now be involved in projects whose capital costs routinely tally into the billions, all monuments to Thomas Alva Edison's vision and ingenuity. If the reader takes only one lesson away from this book—although I hope he or she will take many more—it should be that, no matter what the infrastructure project involved, no project should proceed until its commercial engineering, procurement and construction risks are contemplated and allocated among its participants.

Christopher B. Hansmeyer
Former General Counsel
Abengoa Solar LLC
San Francisco, California

Foreword

Development of large infrastructure projects is a social, economic and scientific adventure. The development and construction process brings together different cultures, professional disciplines and political objectives. A successful project requires intense commitment and careful coordination. A large infrastructure project is a chance to meet and learn the perspective of developers, engineers, political officials, bankers, lawyers, environmentalists, accountants, real estate developers and many others. It is an opportunity to watch and learn how different people and cultures deal with change and the economic benefits and burdens that come with infrastructure development.

Each project has the potential to transform a barren site or untapped water resource into a colossus of steel, concrete and computer chips that can enrich the lives of thousands and become an enterprise that will earn money for decades to come. Development, especially in so-called "developing countries," is an exciting process. However, as an advisor, it is often important to remind clients that there is a difference between being a "developer" who puts together a feasible project under compelling economic circumstances and a "missionary" promoting a project that is politically (and perhaps economically and technologically) ahead of its time.

I must thank everyone who has either put me on the path or kept me on it when I was lost or wandering. It has been an honor to work with each of them. They are far too numerous to mention here (other than my unofficial editors—my Dad and former firm partner J.J. Stevenson).

About the Author

 Howard M. Steinberg is a retired partner and now "Of Counsel" to the project development and finance group of the international law firm of Shearman & Sterling LLP. He concentrates on transactions involving the energy sector and focuses on the power industry in particular. He has represented sponsors, offtakers, fuel suppliers, contractors, operators, underwriters and lenders in the development, acquisition, restructuring and privatization of infrastructure projects around the world. He has had a chapter on Brazil featured in the book *The Principles of Project Finance* and articles in periodicals such as *Power*, *Astronomy*, the Institute for Energy Law's *Energy Law Advisor* and *Asia Law & Practice* and is listed as a leading lawyer in project finance by Chambers Latin America and IFLR 1000.

Mr. Steinberg is a member of the New York State Bar, a member of the Advisory Board of the Institute for Energy Law based in Plano, Texas and holds a J.D. from New York University School of Law, an M.B.A. from Columbia Business School and an A.B. from Columbia College, where he majored in philosophy and economics and graduated *magna cum laude* and *Phi Beta Kappa*.

Introduction

Large infrastructure projects are complex and risky ventures. They are subject to the vagaries of nature, politics, war, capital markets, technological advance, classical economic theory and human error. A large infrastructure project such as a hydroelectric power plant can often take the better part of a decade from the time a developer, utility or government makes a decision to develop a project until the time electrons are being agitated in its generators. It is probably fair to say that most infrastructure projects are never built. They crumble in the evaluation stage, usually for financial or political (not technological) reasons. Of those that make it through the feasibility stage, even fewer make it to construction, generally because they are unable to withstand the great expense and delay of permitting battles over location and environmental effects.

In the case of the electricity industry, by the end of the twentieth century, roughly 100 years after Thomas Edison's Pearl Street power station in lower Manhattan began its operations, a combination of ridiculously optimistic expectations for high power prices, huge projected electricity demand, self-serving assumptions for low fuel prices, easy access to capital, untutored construction cost estimates and pure greed allowed many unfeasible power projects in the United States to withstand the feasibility wrecking ball. The price paid for all this irresponsible financial analysis was high. Many newly constructed power projects sat idle, half-constructed projects lay abandoned and around the world hundreds of millions of dollars of warehoused gas turbines slowly corroded in their packing crates turned coffins. Poor planning led to the economic collapse of the U.S. power industry, bringing down developers, owners, utilities, lenders and contractors in a wave of bankruptcies, lawsuits and buyouts. Beginning in the summer of 2000, the behemoths of the industry, one by one, began to collapse under the weight of their own foolish investments. Wall Street turned its back on the industry. Capital investment and upgrade programs were severely slashed. In August 2003, a blackout affecting approximately 50 million people occurred in the Northeastern United States and Ontario, causing losses estimated at $10 billion by some accounts.[1] Although the dust has settled on this chapter of improvident investment in the United States, 2016 signaled the downfall of many giant international solar power sponsors and "yieldcos" whose stock prices soared based upon planned projects that had not been properly screened for commercial feasibility. Forgetting the fundamentals of project evaluation and management wiped out a mountain of unemployed capital.

Infrastructure project development in the twenty-first century centers on so-called "renewable" energy resources and responsible and realistic construction programs.

No construction program can be devised without a good understanding of the risks involved and the practical and legal allocation of these risks. Identifying these risks is fairly easy. They are the risks and challenges inherent in all construction projects, the same risks that the Pharaohs faced when building the pyramids. What may be more complicated is the contractual relationship and financing involved in these projects and understanding their myriad variations.

Note

1 U.S.–Canada Power System Outage Task Force, Final Report on the August 14, 2003 Blackout in the United States and Canada: Causes and Recommendations (USCBR), 2004.

Chapter 1

A Historical Perspective

Varying Construction Approaches in the U.S. Power Industry

Simply put, a power plant consists of a turbine in which stages of blades are attached to one end of a shaft to cause the shaft to rotate as the result of either steam, water or heat passing through them. The other end of the shaft turns in the middle of a gigantic magnet (known as the stator). Any time metal (the rotator in this case) passes through a magnetic loop (of the stator), an electric current (measured in amperes) will be created as electrons from the magnet are "excited" and bounce into one another. The generator is attached to a "step up" transformer that increases the voltage of the electricity created so that it can travel long distances without unnecessarily losing too much energy. At the location where the electricity is needed, other transformers will then lower the voltage again. The electrons themselves do not actually travel long distances through the wires but simply "bump" into their neighboring electrons and this, in effect, sets off a chain reaction. This "direct current" can travel long distances effectively without losing energy in the power transmission wire (so-called "line losses"). Alternating current (AC), on which most household devices operate, is created by placing metal "wings" on the generator shaft that rotate in and out of the magnet and these "wings" have the effect of allowing the electrons that are excited to migrate back toward their initial position when the "wing" leaves the magnet before the electrons are "excited" again by the next "wing" that swings through the magnet. (Some older buildings [such as Columbia University's Pupin Hall physics building] are still wired for direct current.) By the 1880s, electricity from power plants was illuminating theaters from Santiago to Milan. By 1956, the world's first nuclear power plant in Seascale, England, had entered commercial service.

Traditionally, a U.S. electricity utility that wanted to build a power plant would develop plans and specifications for the power plant that would meet its needs. Next, the utility would purchase the major equipment for the plant from equipment suppliers. Finally, the utility would hire one or more general contractors to build the plant and assemble the equipment according to the utility's plans and specifications. During each phase of this process, competitive bids could be sought and evaluated and the utility could manage and monitor the entire process. This paradigm of the "plan and spec" approach (together with its variants of "design/build" and "construction management" in which the utility owner only undertook preliminary design and then turned the preliminary design criteria over to a firm that would finish the design and build the plant) was prevalent until deregulation of the U.S. electricity industry began in the late 1970s. See Figure 1.1.

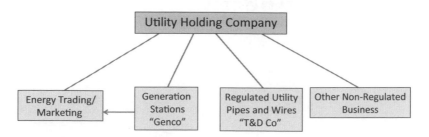

Figure 1.1 Typical U.S. electric utility structure.

The cornerstone of the utility industry in the United States is reliability and safety. With utilities as monopolies without competition, cost and target in-service dates may have been less of a concern to them before deregulation because utilities had legions of accountants and lawyers who could justify their construction and operation costs. Consequently, most utilities could usually rest assured that most (if not all) of their costs would be "passed through" to their customers under their electricity and gas tariffs. A profit on top of cost was usually paid to the utilities as well. In fact, utilities often received a profit on the capital that they had invested in plants under construction and even before they were placed into service. The traditional general contracting approach gave a utility the most control over all aspects of a plant's design and construction and ensured that the utility built exactly what it needed—a reliable generation station with redundancy and wide safety margins. Overruns, inefficiencies and delays were often consequences for the ratepayers to bear through increased electricity tariffs.[1]

The Public Utilities Regulatory Policies Act of 1978 (PURPA) had a significant effect on these contracting customs in the U.S. electricity industry. In an attempt to reduce the United States's dependence on oil and increase efficiency in the electricity industry, Congress encouraged development of so-called "independent power producers" (IPPs) by requiring utilities to buy electricity from independently owned generation stations (so-called "qualifying facilities" or "QFs") if these QFs could meet certain criteria for efficiency of converting fuel into electricity and/or steam (and the steam then needed to be used in another industrial application [so-called "cogeneration"]). A power plant could also qualify as a QF if it used certain types of fuel (such as "biomass" or river water or wind) to create electricity. PURPA required a utility to purchase electricity generated by QFs at the utility's "avoided costs" (that is, the costs that the utility had avoided by not building the power plant itself).[2]

The EPC Approach Emerges

The cottage industry of independent developers who contracted with utilities to sell power to them under the PURPA regime needed capital to build their plants. In order to access this capital, developers had to demonstrate that they could build their plants on budget (so that no unexpected capital would be needed), on schedule (so that the "drop-dead" dates for their plants achieving commercial operations under their utility power purchase agreements would not be triggered), and at the proper efficiency and output levels so that their regulatory status as QFs and long-term profitability would not be jeopardized. These independent power producers did not have the experience or financial stature to deal with these risks. They turned to large engineering and construction companies to take these risks for them. These companies were willing, but they sought a hefty premium above the traditional general contracting approach in return for taking such risks. They became known as "turnkey EPC contractors" because they would handle all aspects of engineering, procurement and construction for a fixed price. Thus, a project sponsor could enter into a "turnkey, lump sum" engineering, procurement and construction (EPC) contract under which an EPC contractor would agree to construct an entire power plant by a fixed date in the future. If the EPC contractor did not construct a plant in which the sponsor could "turn the key" and start operations by the agreed-upon date, the EPC contractor would be obligated to pay daily liquidated damages (up to some capped amount—usually approaching 15 or 20 percent of the EPC contract price) for the delay. Most often, the amount of these liquidated damages is set at a level that roughly compensates the developer for its costs associated with not having the plant in operation by the "guaranteed" agreed-upon date. Ideally, these liquidated damages will cover the interest payments required to be made to the sponsor's lenders plus any liquidated damage payments the sponsor will have to pay to the power purchaser for the generation station not being operable by the date that the sponsor promised the power purchaser in the power purchase agreement plus the profit the sponsor expects to lose as a result of not being able to sell power to the power purchaser beginning upon the agreed-upon date. In practice, the liquidated damages that a responsible EPC contractor will agree to pay are often significantly lower than all the costs listed above.

In addition, EPC contractors pay additional liquidated damages if the power station does not meet certain minimum performance levels in terms of electrical output potential (often referred to as electrical generation "capacity" or "capability") and efficiency in turning fuel into electricity (known as "heat rate"). If a sponsor engages an EPC contractor to design and construct a power plant capable of producing 1,000,000 watts (a megawatt [1 MW]) and, when built, the plant can only produce 999,000 watts at any given time, the sponsor will lose the ability to sell the shortfall of 1,000 watts (a kilowatt [1 kW]) to its customer (the power purchaser) for the life of the power plant (perhaps 20 years or more). Similarly, if the sponsor contracts with the EPC contractor to design and build a power plant that requires 8,000 British thermal units (Btu) of heat energy to create 1,000 watts of electricity in an hour (a kilowatt hour [1 kWh]) but the EPC contractor actually constructs a power plant that requires 9,000 Btu of heat energy to create 1,000 watt hours of electricity, the power plant will always need to consume more fuel to produce electricity than the sponsor has originally forecast. These "performance shortfall" liquidated damages (usually capped at 15 percent of the EPC contract price in total or sometimes individually at 5 or 10 percent for each of heat rate and capacity, often called "subcaps") help compensate a sponsor for its prospective financial opportunity costs and losses. Obviously, both of the above scenarios can have a serious impact on a project's profitability and even solvency. In fact, "performance shortfall" liquidated damages, because they are capped and only estimates of future losses, generally will not ever fully compensate a sponsor for the losses that are likely to be incurred as a result of shortfalls in performance (as opposed to construction delays, which are usually of less consequence unless the delay is exceedingly long and not covered by delayed startup insurance—see Chapter 23).

The Collapse of the U.S. Power Industry

Owing to careful planning and responsible participants, many power projects of the 1980s and 1990s were very successful. However, as competition in the U.S. power industry increased throughout the 1990s and sponsors began developing new "merchant" plants (which had their future electricity production not committed to be sold under long-term power purchase agreements but instead to be sold into the electricity spot market), financial managers and traditional utility construction managers became reluctant to pay the risk premium EPC contractors commanded. Sponsors began to assume more and more of the engineering and construction themselves or turned to inexperienced EPC contractors who were willing to undercut their competition. Capital investors, following a decade-long streak of successful projects, relaxed their standards of project scrutiny. EPC contractors, attempting to retain market share, often matched the pricing of new and inexperienced EPC competitors. Increased construction activity led to increased competition for major equipment such as gas turbines. Large sponsors began to borrow money to order gas turbines from vendors without even having identified projects in which to install these turbines. Banks began to lend money to portfolios of projects instead of single projects (including projects still on the drawing board). Tremors in electricity prices and fuel prices were all that were necessary to knock out the sagging supports of the industry's arrogant approach to the fundamentals of good construction budgeting. The U.S. electricity industry was soon buried alive in its own rubble as ill-conceived projects were unable to shoulder the billions of dollars of debt resting on their hastily-assembled financial structures. Sponsors abandoned projects and many of the sponsors that had obtained non-recourse project financing left their lending banks with the useless collateral of half-built projects and stockpiled gas turbines. The heap of fiscal and physical waste was carted away by the wrecking crews of accountants, law firms, salvage contractors, arbitrators and judges. As a result of these bad investments of tragic proportions, project sponsors around the world have returned to focusing on the contractual foundation of project development so that these reinforced project arrangements allow capital investment in infrastructure to once again bear the great weight of the risks involved in construction projects. The next chapters will examine these risks.

Notes

1 Electricity utilities have historically been highly regulated by both the federal government and state public utility commissions. State regulatory authorities generally regulate almost all aspects of a utility's retail operations. The principal federal legislation that governs public utilities is the Federal Power Act of 1935 (FPA). The Federal Energy Regulatory Commission (FERC) regulates wholesale sales of electric power and the transmission of electricity in interstate commerce pursuant to the FPA. FERC also subjects public utilities to rate and tariff regulation and accounting and reporting requirements, as well as oversight of mergers and acquisitions, securities issuances and dispositions of facilities.

2 The Energy Policy Act of 1992 (EPA) engendered further competition in the electricity industry. Among other things, the EPA created a new class of power stations known as exempt wholesale generators (EWGs). In order to obtain status as an EWG, a power generation station must be exclusively engaged in the business of selling electric energy at wholesale. An EWG is considered a public utility under the FPA and is subject to FERC regulation of its power sales. With the Energy Policy Act of 2005 (EPAct 2005), Congress scaled back utility commitments to purchase power from QFs located within Regional Transmission Organizations, where it is thought that IPPs are offered better conditions for selling their service competitively. Also, with EPAct 2005, Congress repealed the Public Utility Holding Company Act of 1935, smoothing the way for utility consolidation, and relieving IPPs of many of the restrictions that EWG status was designed to avoid.

Chapter 2

Choosing a Development Approach

A sponsor's approach to the construction of an infrastructure project will depend upon its experience in construction, its tolerance for financial risk and schedule delay and its source of project funds. A well-capitalized sponsor with extensive construction experience in the location of its project may be best served by pursuing the design and construction management approach of letting contracts for discrete portions of the work separately to different contractors in order to squeeze the last dollar of savings and efficiency out of the overall capital cost of its project. On the other hand, a sponsor without recent and successful experience in project construction in the vicinity of its proposed project that is planning to solicit third-party funding to help finance its project is probably better served by pursuing the turnkey, lump sum, EPC approach.

In general, investors of equity and lenders of debt have very limited tolerance for the risks involved in the design and construction management approach to a project, usually as a result of their desire to be only passively involved in the project (and also, if they are lenders, of their limited ability to share in the "upside" of the successful management of risks). Lenders, aside from "upfront" and "agency" fees, never receive more than the interest rate on their loans (unless they have negotiated the right to purchase equity in the sponsor's project or otherwise exchange their loans into equity in the sponsor's project in order to share in the "upside," which is not very typical of lenders who are financial institutions or banks). Thus, in most cases, capital restrictions and capital accessibility dictate that the EPC approach to a project be used.

Concessions

While many projects are truly private commercial transactions (for example, a sponsor constructing a power plant to serve a privately owned aluminum smelter), many infrastructure projects are essentially public works projects (for example, a sponsor constructing a hydroelectric dam to serve local electricity and water needs). Concessions (in which a private enterprise [the sponsor] undertakes a project for the public good) can take many forms, but most typically the government, or one of its agencies, will execute an implementation agreement that outlines the terms of the concession and the instances in which the concession may be revoked. Concessions are often awarded because a government does not want to expend the funds required or assume the risks involved in designing and constructing a project, and instead the government can pay for the capital cost of the project over time by finding a concessionaire to undertake the project. Of course, the concessionaire will expect a profit in addition to its capital investment in the project. Thus, since concessions are likely to cost the government and its citizens more money, the government must balance the benefit of retaining the flexibility to employ its capital in other areas against investing in a project that could instead be "let" as a concession (Volume II, Part E contains a "checklist" of typical concession agreement topics). Some concession approaches are outlined below. See Figure 2.1.

Build and Transfer (BT)

In this case, a sponsor (or EPC contractor) agrees upon a price to build a project for the government.

Build, Transfer and Operate (BTO)

In this case, a sponsor builds a project and transfers it to the government, but then continues to operate the project for the government.

Build, Own, Operate and Transfer (BOOT)

In this case, a sponsor builds and operates the project for a significant period in which to earn a fair return on its investment and then transfers ownership and operation of the project to the government.

Build, Own and Operate (BOO)

In this case, a sponsor constructs, owns and operates the project for its useful life.

Public–Private Partnerships (PPP)

In this case, a sponsor enters into some type of alliance with a government to develop and operate a project with the government as its "partner." Generally, the sponsor will have operational control of the project under this scheme but the government will remain as a passive partner to monitor operation and performance. These projects are particularly useful for a government that wants to gain experience in a particular area (such as high-speed railroads) from a knowledgeable concessionaire.

Privatizations

In the case of privatization, the government transfers ownership of an existing facility (such as a bridge or port) to a private sponsor (often through a sale intended to raise money) and the private sponsor takes over operation of the facility (and often commits to upgrade or expand operations with the goal of increasing the efficiency and profitability of the facility). Occasionally the government will maintain some level of control over the profitability of the privatized entity so the private sponsor is not able to reap windfall profits at the expense of consumers, either directly by means of profit sharing or indirectly by means of rate setting regarding the facility's services.

Legal Regimes—Civil Codes vs. Case Law Approaches

In "civil code" jurisdictions, codified legal principles govern commercial contracts and courts rule upon how to apply the code in the particular case at hand. Their reasoning for their decision is often not even explained in writing and is of no value in another court case. Thus contracting parties are left to the civil code for guidance on its provisions. In so-called "common law" jurisdictions, judges often "report" their opinions on cases in writing and their decisions can be used for precedential value in future court cases if they are similar. Given the intricacies of the EPC approach, it tends to be more common that the parties will select a common law jurisdiction to govern their EPC contract because they can look to case law to help determine how the terms of their contract will be viewed by a court. Most jurisdictions permit parties to choose any jurisdiction to govern a contract between them.

Choosing a Legal Entity

To maximize profits and reduce risks, a sponsor is best advised to set up a legal structure that will help achieve the foregoing goals. In most jurisdictions around the world there are different forms of legal entities under which a business can operate.

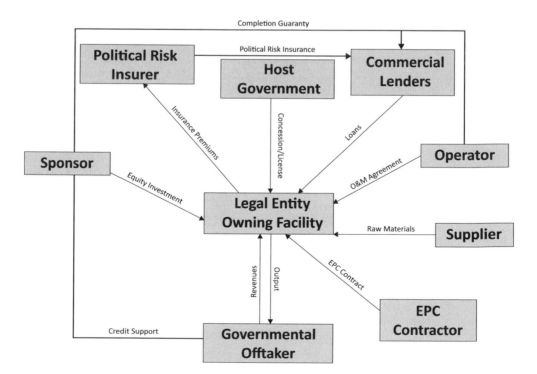

Figure 2.1 Typical project concession.

General Partnerships

In the United States, aside from a sole proprietorship, the simplest form of enterprise is a partnership in which two or more people or legal entities (such as corporations) form a partnership to carry out their business. Each partner has unlimited liability for the obligations of the business. Thus, each partner's assets are at risk to satisfy the obligations of the partnership. To the extent the partner has assets, these assets are at risk of being subject to the claims of creditors of the partnership. This type of entity is known as a "general partnership" in the United States and generally the partners are free to agree upon any type of profit sharing and governance arrangement that they desire. If they have not agreed upon a matter and a dispute arises between them, generally a state statute or common case law may dictate how such a matter is to be resolved. In a general partnership, any partner has the legal power to bind the general partnership to any obligation. No legal filing is necessary to create a general partnership although it is a legal entity distinct from its owners. (In other jurisdictions, these are often known as *Offene Handelsgesellschaft* [OHG; Germany], *société en nom collectif* [France] and *sociedad regular colectiva* [Spain].) In the United States, a general partnership is not treated as a federal income-tax paying entity (but may be treated as a tax payor by state taxing authorities, so it is important to determine this before a state of incorporation is selected). The partners pay U.S. federal income tax but not the partnership. The partnership, at the end of each fiscal year, will send its partners an Internal Revenue Service schedule K1 (known as a "K1") indicating the tax (or loss) allocable to each partner, so each one can calculate its own tax liability accordingly. Partners are also able to make "special" allocations of profits and losses between one another if there is a valid economic reason for this allocation, which can provide further benefits for their economic return on a project.

Limited Partnerships

Limited partnerships are the same as general partnerships with one exception. Limited partnerships contain a special class of partners known as limited partners who do not actively participate in the management of the limited partnership (but often are granted veto rights over fundamental matters). As a result of this "non-participation," the benefit of "limited" liability with respect to the limited partnership's creditors is bestowed upon these limited partners. Limited partners are not liable for the debts of the limited partnership (of course, they can lose the money they have invested in the limited partnership, but that is all). The state law often makes an exception to this rule. If the limited partner has been actively engaged in the management of the limited partnership, then it will become a general partner and become personally liable for the debts of the limited partnership. All limited partnerships must have at least one "general" partner who is responsible for managing the limited partnership's business. This general partner, as in a general partnership, has unlimited liability for the debts of the partnership. (Based on the same principles as limited partnerships are *Kommanditgesellschaft* [KG] in Germany, *société en commandite par actions* in France, and *sociedad comanditaria* in Spain.) Limited partnerships generally file "certificates of limited partnership" with the secretary of their state to create their existence and copies of this certificate are available to the public. Limited partnerships, like general partnerships, do not pay U.S. federal income tax.

LLPs and PCs

In certain jurisdictions in the United States, professionals such as doctors, lawyers, engineers and accountants may form professional limited liability partnerships or corporations (often denoted by "LLP" or "PC" or "professional corporation"). These entities allow the professionals who are the owners to insulate themselves against typical commercial liabilities such as to a landlord or phone company but not for their own professional malpractice or that of their "partners" or associates who work with or for them. This is why they often carry professional liability insurance (see Chapter 23).

Corporations

In the United States, corporations are often referred to as "fictitious legal entities" in law school textbooks because corporations are empowered to act as persons under the law separate and apart from their owners. The word "corporation" is derived from the English translation of the Latin word *universitas*. In fact, the first corporations were universities. A corporation has stockholders who contribute their money to the corporation and then elect a board of directors to oversee the business of the corporation. The board of directors appoints officers to carry out and execute the business of the corporation. All of the procedures and rules governing the issuance of stock and the election and powers of the board of directors are generally prescribed by statute, and usually may not be altered by stockholders or directors. Stockholders have no liability for the debts of the corporation except in very rare circumstances when the "corporate veil" insulating the stockholders from the corporation's creditors is "pierced" (such as when the government pursues an environmental claim under a federal statute, or the stockholders have engaged in fraud to deceive creditors by disposing of the corporation's assets without receiving fair value for them [a so-called "fraudulent conveyance"]).[1] A corporation is established by filing articles or a certificate of incorporation with the state secretary and copies thereof are available to the public. (In other jurisdictions corporations are known as *Aktiengesellschaft* [AG; Germany], *société anonyme* [S.A.; France) and *sociedad anónima* [SA, Spain].) Corporations must pay U.S. federal corporate income tax.

Limited Liability Companies

Limited liability companies are a hybrid of partnerships and corporations. They are a fairly recent creation in the United States and were designed to blend the most favorable attributes of partnerships and corporations while preserving the U.S. federal non-taxable nature of partnerships. Limited liability companies are comprised of members who are not liable for the debts of the limited liability company and who may appoint one or more members (or other individuals or entities) as the manager or managers of the business of the limited liability company. Statutory provisions generally govern the administration of limited liability companies but members are usually free under their state statute to alter these statutory provisions. Limited liability companies are formed by filing a certificate with the state secretary, and copies thereof are available to the public. (Limited liability companies are similar to *Gesellschaft mit beschrankter Haftung* [GmbH] in Germany, *société à responsabilité limitée* [SARL] in France, and *sociedad de responsabilidad limitada* [SL] in Spain.) As with general partnerships and limited partnerships, limited liability companies do not pay U.S. federal income tax, but their members do.

Forming the Project Entity

Most sponsors want to preserve all their options and their own existence. For instance, many sponsors or utilities are publicly held corporations (which means that their stock is available to be purchased by the public on a securities exchange or over a trading system) and have significant assets. Their plan is often to develop and then sell some percentage (or all) of each new project. Since they usually have extensive assets, they are probably not best served by entering into contracts and loans for a new project in their own name and exposing all of their assets to the obligations and potential claims under these contracts. Generally, a sponsor or utility (or even a government) will form a new subsidiary company to develop, own and operate a new project (commonly referred to as the "project company"). There are many reasons for this. The foremost is flexibility. Ownership through a subsidiary will allow the sponsor to find other equity investors or partners who are willing to make passive or active investments in the sponsor's project (see Figure 2.2). It will allow the sponsor to sell the entire project easily (including transferring the project's permits and real estate) at any time by simply selling the stock or partnership interests in the subsidiary to a buyer without the unnecessary complications of deeds and assignment instruments.

From an accounting standpoint, a further advantage of creating a subsidiary is that the sponsor usually need not consolidate the debt of any of its subsidiaries on its own financial statements unless the sponsor owns more than 50 percent of that subsidiary or otherwise controls that subsidiary in another way, such as through a management contract.

Finally, should a subsidiary become insolvent during construction or operation because its project has problems, the subsidiary may seek the protection of the U.S. Federal Bankruptcy Code (discussed in Chapter 13), often without the risk of pulling the sponsor itself into bankruptcy.

If a subsidiary is not set up to own the project, a buyer would have to buy an interest in the sponsor or utility itself or purchase each item of the project separately from the sponsor or utility if it wanted to participate in the project. An advantage of this approach is that the sponsor can isolate the financial risk associated with a project by investing only as much capital in its subsidiary as is necessary to carry out the project. Generally, the creditors of the subsidiary, such as the EPC contractor, the fuel supplier and the lenders, will not have any recourse to any party other than the subsidiary to satisfy their claims because they will be conducting business with (and entering into contracts with) the subsidiary and not the sponsor.

SPVs

The business of a company can be limited exclusively to carrying out its project. This limitation can be merely self-imposed by the sponsor, contractually agreed by the company (and its owners) with any or all of the company's contractual counterparties, or even stated in the certificate of formation of the company filed with the state secretary. Such a company whose business has been restricted to a specific purpose is often referred to as a "special purpose vehicle" (SPV), "special purpose company" (SPC), "special purpose entity" (SPE), "Project SPV," or "project company" but such an entity has no special legal status as an entity. Thus, an SPV is not a type of legal entity but rather just a term of art used as shorthand for a company whose business has in some way been restricted.

Of course, if an SPV is to be used, the lenders and suppliers are sophisticated parties and will understand they are contracting with a "shell" SPV without assets other than a (to be constructed) infrastructure project. Thus, lenders will likely charge a premium to lend money to an SPV and demand a pledge of all of the SPV's assets (including its equipment and real estate) in order to secure their loans to the SPV.[2] Lenders will also require the sponsor to agree that the sponsor will capitalize its SPV at some minimum amount of equity (at least until the SPV's project is commercially operable; at which time lenders will often permit the sponsor to withdraw cash from the SPV on a quarterly or semi-annual basis as long as such cash is not needed for the operation or maintenance of the SPV or to make payments on the SPV's debt, and the SPV has been able to maintain a satisfactory ratio of its cash flow to its payable debt service [referred to as the "coverage ratio", see Chapter 4]). This release of cash is important to the sponsor, because cash unnecessarily trapped inside an SPV is cash that cannot be invested in developing other projects or returned to the sponsor's owners.

In addition, suppliers will usually limit their credit exposure to an SPV by agreeing to sell only a certain level of fuel or equipment on open account to it. Sometimes, suppliers will seek payment security in the form of cash escrows or letters of credit from the SPV. In fact, in the current market, even offtakers of an SPV typically ask it for credit support to ensure that the SPV will honor its delivery obligations (often in the form of letters of credit or even a second lien or mortgage on the SPV's equipment and real estate). This is a typical arrangement, especially in the case of power plants, because power purchasers often have obligations to deliver power to other utilities or their own retail consumers and can face great expense (or even fines) if they fail to deliver power to their own customers during times of high power demand.

Development Partners

There can be many advantages for a sponsor in finding partners in its project company to assist in the development of the project. Of course, a sponsor that wishes to include a partner in its project must evaluate whether or not the advantages of having a partner in its project company justify the administrative, reporting and bureaucratic complications that having a partner in its project company will create. When a sponsor is involved in a project outside of its home jurisdiction, a local partner in its project company may be particularly helpful because it will often be conversant with

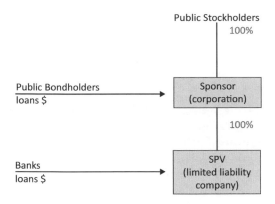

Figure 2.2 Typical sponsor ownership structure.

local practices and customs (or even required to be by the local law). Sometimes, given a project's cost or complexity, a sponsor might want a partner with access to funding (such as a venture capital fund) or technical expertise (such as a mining company). In other situations, a sponsor might choose an EPC contractor or equipment supplier as a partner in its project company in the hopes of getting better pricing on the project's EPC contract or equipment, and comfort that during operations the EPC contractor's investment and expertise will lead to better operating performance (this may be especially true in the case that the sponsor is going to employ new and untested technology in its project from a vendor). The sponsor must be cognizant that conflicts of interest can arise if a partner desires to have one of its affiliates provide services to the project. The sponsor should protect itself against the consequences of conflicts of interest by including protections in the organization's documents of the project company that will govern the relationships among the parties (that is, the shareholders' agreement in the case of a corporation, the operating agreement in the case of a limited liability company, or the partnership agreement in the case of a partnership). At a minimum, these protective provisions should include a requirement for fair dealings in all contracts between the project company and any parties that are affiliated with any of the owners of the project company.

As a practical matter, affiliates of an owner in the project company that wish to provide services to the project should probably be given only a right of "first offer" to provide services or equipment to the project. The project company should be able to refuse these services or equipment so long as, in the interest of fairness, there is an understanding between the affiliate and the project company that the project company will not subsequently enter into any arrangement with another provider on terms less favorable to the project company than those that were offered to the project company by the affiliate. However, an affiliate of an owner should not be given a right of "first refusal" (as opposed to a "right of first offer") to top any other service or equipment provider's offer, because a right of first refusal is likely to discourage other providers from taking the time to make competitive offers if they know that they can be "knocked out of the running" at the last minute by an affiliate of an owner of the project company. Other service or equipment providers will generally not be interested in being used by the developer as a stalking horse so that the project company can obtain the lowest price from an affiliate of one of its owners in the project.

If a project company enters into a partnership with an EPC contractor, the sponsor must understand that, unlike the sponsor itself, the EPC contractor will typically be expecting to make most of its profit from the payment it receives under the EPC contract to build the project and not from the project's expected revenues. In fact, it is likely that the EPC contractor will be interested in selling all of its equity interest in the project company as quickly as possible once the project is complete (or even sooner) so that the EPC contractor can move its capital to other projects that the EPC contractor has in its development pipeline. Thus, the sponsor, if it is relying on the EPC contractor's experience to help guide the project through its nascent operational phase, should require the EPC contractor to maintain its equity investment in the project for at least a "shakeout" period after the project enters commercial operation. However, even before operations begin, conflicts of interest can arise if changes must be made in the work (discussed in Chapter 12) and the EPC contractor seeks more money or time to perform these changes. Thus, it is important that the sponsor makes sure that the EPC contractor, when acting in its capacity as one of the owners of the project, does not have the right to vote on matters regarding the amendment or enforcement of the EPC contract. In fact, from the sponsor's point of view, it is best that the sponsor maintain as much control as possible over its project and confer on its partners the right to vote only in certain key matters concerning the project (so-called "super-majority" rights that are granted in order to protect minority partners against fundamental organizational and business changes regarding the project).[3]

An Integrated Structure

Most project structure centers around an SPV (for example, see Figure 2.3) through which all funds will flow and from which all services will originate. This SPV "hub" is only as good as its contractual spokes. Without a balanced and galvanized contractual risk allocation mechanism for each contractual spoke, the project's wheel of life can quickly become out-of-round and off track as it rolls down the perilous road of project development. The following chapters will try to expose just how treacherous the road can be.

Notes

1 See *Port Chester Elec. Constr. Corp. v. Atlas*, 40 N.Y.2d 652 (1976) (refusing an electrical subcontractor's attempt to "pierce" the corporate "veil").
2 In fact, for lenders, the first rule of lending is to lend to an entity that owns assets and not an entity that owns "paper assets" such as stock certificates of another entity.
3 See *Okland Oil Co. v. Knight*, 2003 WL 22963108 (10th Cir. Dec. 17, 2003), for a good discussion of some of the issues that can arise when an equity investor in a project also has an equity interest in the project's EPC contractor.

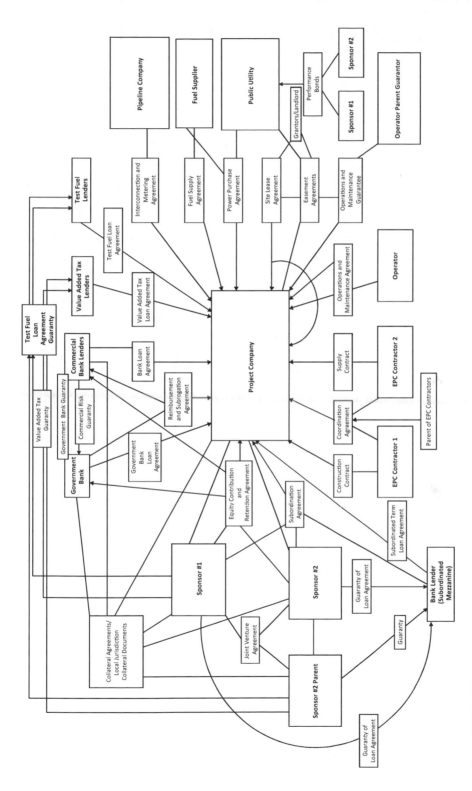

Figure 2.3 Typical power project structure.

Chapter 3

The EPC Contractor's Approach

Construction contracting, and EPC contracting in particular, is a risky business. EPC contractors must be well prepared for their undertaking and will generally find little sympathy in U.S. courts for any economic or construction difficulties they encounter. The U.S. judiciary generally takes the view that a contractor was free to protect itself against any risk that may have arisen to cause it harm by allocating that risk to the sponsor in its contract and, if the sponsor would not agree to accept the risk, the contractor was under no legal compulsion to execute the contract. As one judge somewhat ruthlessly opined:

> The work to be performed is grounded in the principles of free enterprise and competition. The scope of such an undertaking requires those who accept the invitations to bid to be well versed, or advised, in contract law and forms together with comparable efficiency in accounting and engineering departments. Otherwise their financial doom is spelled. It is common knowledge that the owner, or sponsor, "writes its own ticket" and the competing market of contractors is obliged to bid and perform in conformity thereto, or to avoid that field of endeavor.[1]

Even with this high level of risk, EPC contractors generally do not set up special purpose subsidiaries for each "job" or EPC contract that they perform. Often, EPC contractors are not publicly held companies but are "closely held" companies whose stock is owned by a few individuals and not listed for trading on any stock exchange or automated quotation system. Large EPC contractors will often have many operating subsidiaries, each of which performs different functions or operates in particular geographic regions. It is important for the sponsor to understand which of the EPC contractor's subsidiaries will sign the EPC contract and whether that subsidiary can operate on its own and independently from its parent entity should its parent entity run into a cash flow problem (or worse, bankruptcy). Under legal principles in most U.S. jurisdictions, parent companies are not liable or responsible in any way for the debts or obligations of their subsidiaries unless they so agree in writing or have acted with fraudulent intentions. This is true even if the parent entity has received a benefit from its subsidiary's entering into the EPC contract.[2] In order to avoid the unfortunate consequence of contracting with a subsidiary that does not have the technical expertise and financial stability to perform its contract, an unconditional guaranty of performance of the subsidiary is typically obtained from its ultimate parent (as will be discussed in Chapter 19).

Consortia

When large or risky projects are involved, EPC contractors will often form what they will refer to as a "consortium" or "joint venture" or "alliance" with other EPC contractors to, in their words, "strengthen" their commitment to the sponsor's project. This strategy enables them to share risk and usually improves the chances for budgetary success in a complicated project in which many different areas of expertise are required. It is important to know the legal form that such a collective effort will take.

Usually, new entities (without any substantial assets) will be formed to make up the consortium and, therefore, it is crucial for the sponsor to insist on guaranties from the ultimate parent entities of the consortium members. Frequently, this consortium approach is to the sponsor's benefit because the sponsor's SPV often will receive guaranties from two or more unaffiliated EPC contractors, thus decreasing the SPV's exposure of having only one party guarantying the performance of the legal entity that will sign the EPC contract.

The consortium generally does not share the agreements governing the relationship between the consortium members (such as a joint venture agreement, which does not even create a legal entity but just a contractual relationship) with the sponsor but may do so if the sponsor asks to see the arrangement. Thus, it is a good idea for a sponsor to ask to see the arrangement because usually the arrangement will allocate the risks that the contractors have identified and that they believe are probable enough that they must be allocated between the contractors. The sponsor may not even have comprehended these risks. Sponsors should be careful that any guaranty received from a parent entity of a consortium member is not limited in any way and that guaranty should not limit a parent entity's responsibility to those actions taken by its own subsidiary but covers any action taken by any member of the consortium. Otherwise, if the sponsor's SPV makes a claim against the consortium, each parent entity could claim that its subsidiary was not responsible for the problem in question and the sponsor's SPV could be stuck in a costly legal spiderweb of counter and cross claims in an attempt to sort out who is the responsible contractor. All this can be avoided by negotiating comprehensive guaranties. Any parent entity that wants to limit its own liability to that created by its own subsidiary can instead do this without involving the sponsor by executing an indemnification agreement between itself and the other parent entities involved in the consortium so that claims by the SPV can be apportioned among the parents.[3] The sponsor must consider whether it will want to negotiate the right (but not the obligation) to combine all proceedings against a parent guarantor with proceedings under the EPC contract so that all disputes can be heard in one forum at the same time; otherwise, a sponsor could possibly be required to pursue claims in different jurisdictions at the same time (see Figure 3.1).

The EPC "Wrap"

In the industry, it is common to hear that a company will "wrap" the EPC contract. Not surprisingly, this is not a legal term. It is simply industry vernacular for the idea that one legal entity will be responsible for bringing together all the contractors to be involved in the project and then warrant the performance of the project in terms of output, efficiency and scheduled in-service date. This is done by one company

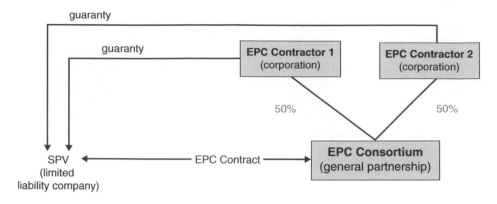

Figure 3.1 Typical EPC consortium structure.

executing an EPC contract and taking full responsibility for a project vis-à-vis the sponsor, and then the company's EPC entering into separate "sub" contracts with all the other contractors involved. Thus, there is one entity to depend upon for performance and this so-called "EPC contractor" is paid a premium for taking the risk of properly consolidating all the "sub" contracts for the project. Although it seems to be a somewhat pejorative connotation, in some sense EPC contractors can be viewed as "aggregators" of engineering services, construction services and equipment. This "wrap" structure is often employed by Japanese trading companies that neither manufacture anything nor perform engineering services. They act essentially as the agents of their other Japanese counterparts that can deliver these products and services but who are not capable of coordinating an entire project or doing business overseas.

On-shore/Off-shore Issues and Coordination Agreements

In many countries, goods and services are taxed at different rates and imported goods and services are also taxed at different rates from domestic goods and services. If one EPC contract for the sale of both goods (equipment) and services (engineering) is executed, local tax authorities may choose to apply the highest of tax rates. With this in mind, and to avoid unnecessary taxation, sponsors often ask EPC contractors to sign two separate contracts, often with one domestic subsidiary of the EPC contractor and one non-domestic subsidiary of the EPC contractor, one for the supply of domestic goods and services and one for the supply of imported goods and services. This addresses unnecessary taxation but it creates a potential problem because now one legal entity is no longer responsible for the entire EPC of the sponsor's project. In order to prevent any goods or services from "falling through the cracks" and creating a technical or mechanical void, a so-called "coordination agreement" is usually executed by the ultimate parent company of the EPC contracting entities that provides that the parent company will be responsible for delivering to the SPV what the SPV bargained for—a fully operable facility. This "split" approach generally benefits all parties and avoids unnecessary taxes, but it is important to address questions of which parties will be responsible for unanticipated taxes (and possibly penalties) if tax authorities successfully raise objections to this type of "split" arrangement and determine to treat the two separate contracts as one contract for taxation assessment purposes.

EPCM

Sometimes contractors and developers talk about the "EPCM" approach as if it was a variation of the EPC approach. Under the EPCM approach a sponsor hires a manager (usually an engineering firm) to supervise and manage engineering, procurement and construction in return for payment for its time, and often a contingency fee as well if the project is completed on time and on budget. (This approach is typical in the mining industry, because mining projects involve many different engineering disciplines and are often in remote locations. Few EPC contractors are able to offer services that can cover all the different project aspects while offering a price that is competitive as compared with a sponsor bidding out separate work scopes and managing the various winning contractors.) EPCM has nothing to do with a lump sum, turnkey EPC contract approach. In the EPCM approach, the construction manager may put some or all of its fees at risk but the sponsor will be financially responsible for almost all delays and overruns. Furthermore, if all project contractors perform satisfactorily in their own right, but their work and interaction has not been properly coordinated by the construction manager, they may be able to escape responsibility for the costs of delay and defects. Such costs can often far exceed the liability of the construction manager to the sponsor, so the sponsor will have to use its own funds to resolve these problems. For instance, in the case of a power plant, if the sponsor was expecting a heat rate of 7,000 Btu/kWh but the plant that was built has an 8,000 Btu/kWh heat rate, it will be the sponsor that must bear the burden of proving which contractor or contractors were at fault for the deficiency if all the contractors deny culpability. Since most design and construction risk will be shouldered by the sponsor and not the EPC contractor (because there really is no EPC contractor) under this approach, lenders are unlikely to agree to non-recourse project financing in which the sponsor places at risk only the assets of the project and a predetermined amount of its equity capital. In fact, in the case of mining projects that are carried out on an EPCM basis, their financing requires generally that the SPV's ultimate parent give the SPV's lenders a so-called "completion" guaranty that provides that until certain completion test requirements are met, the ultimate parent of the SPV will remain liable to guaranty the debt service of the SPV to its lenders. These completion tests often require a certain level of production of a certain quality of product while not exceeding a specified operating cost target level.

Notes

1 *Terminal Constr. Corp. v. Bergen County*, 18 N.J. 294 (1955).
2 See *Intergen N.V. v. Eric F. Grina, Alstom (Switzerland) Limited, and Alstom Power N.V.*, 344 F.3d 134 (1st Cir. 2003), noting that there was no "compelling policy objective" for determining that a parent entity was an alter ego of its subsidiary so that it should be bound to one of the terms of its subsidiary's contractual agreements.
3 See *Siemens Westinghouse Power Corp. v. Dick Corp.*, 293 F. Supp. 2d 336 (S.D.N.Y. 2003), in which consortium members failed to agree upon their apportionment of liability for the payment of liquidated damages to an owner.

Chapter 4

Beginning the Evaluation Process

The Financial Model

Once a development approach and form of entity have been chosen by the sponsor, the next step is to prepare a more thorough forecast of expected revenues and expenses of the project during its operational life (often called a "pro forma projection"). This financial model can be prepared once the sponsor has made some preliminary assumptions about the size of the project, its efficiency, its expected utilization factor (often referred to as "capacity factor" in the power industry [which connotes how many of the 8,760 hours in a year the facility is expected to operate]), its budgeted capital cost (plus some allowance or "contingency" for cost overruns—in most industries between 5 and 10 percent of the construction cost of the project) and the amount, tenor and interest rate to be associated with the portion of the project's capital costs that will be funded by loans or preferred equity. By far the largest operating costs are generally fuel and repayments of the loans (so-called "debt service" payments) or preferred equity capital. These costs, combined with operating and maintenance costs, will determine whether the undertaking can be expected to be profitable.

Whether or not the project will be a profitable investment for its sponsor is dependent upon whether or not it is built within its budgeted cost. Unexpected cost overruns can quickly turn an anticipated double-digit internal rate of return on a risky project into a single-digit rate of return on a par with the most risk-free United States Treasury bill that could have been purchased as an investment instead of building the project. It is crucial to make an informed estimate of the capital cost to be used to evaluate economic feasibility and then perform a "break even" analysis to determine at what point the sponsor's capital investment is not likely to ever be recovered because the project will not be able to generate enough revenue to repay its own capital cost. Initial capital cost estimates should be constantly updated as the development process progresses.

Financing

"Project financing" has developed because infrastructure projects generally create a fairly reliable and predictable stream of long-term revenue. So long as a project has long-term offtake and feedstock contracts at calculable prices or has a fixed payment based on the availability for service of the project and its counterparties are creditworthy, once construction is completed, there remain only operational risks (such as poor maintenance or management) and these risks will be minimized if a responsible and experienced operator is going to operate the project.[1]

As was discussed in Chapter 2, in return for paying lenders a "premium" or "spread" above a reference interest rate, owners are generally able to put a set amount of capital at risk and borrow the remaining capital necessary to build their project. Lenders typically will lend between 50 and 80 percent of projected capital cost (including money to pay the lenders their interest that accrues during construction, so-called "IDC") required to build a project on a "non-recourse" or "limited recourse" basis. This means that once the sponsor has made, or is committed to make, a certain level of capital contributions to the SPV, lenders will look only to the SPV and its assets (the SPV's facility, its contracts and its receipts) to satisfy their debt if a project slides into financial difficulty.

It is usually possible for a sponsor to improve its internal rate of return on a project by postponing for as long as possible the equity investments it must from time to time make in a project to fund costs, especially if the project has a lengthy construction period. Thus, in the case of a project with a three-year construction cycle, if the SPV borrows funds from its lenders for the first two years of construction and then the SPV's owner begins to make equity contributions to fund construction thereafter, the postponement of the owner's investment in the SPV will enhance the project economics. Of course, the lenders will want a written commitment from the sponsor that the sponsor will invest these funds when the funds are required and they may insist upon a letter of credit from the sponsor to "backstop" this investment obligation. The sponsor will have to take into account the costs of posting this letter of credit (see Chapter 19).

A large project may have many different tranches of loans with different purposes and tenors. Local lenders may provide loans in local currency to fund local costs and value-added taxes. Other lenders may fund loans in the currency of the countries in which major equipment suppliers are located. Some lenders desire to make only relatively short-term loans, while others are prepared to fund longer-term loans. It is the owner's task to make certain that payments to be received from the SPV's customers will match all these different currencies, maturities and interest rates.

It is probably fair to say that most projects become distressed as a result of unanticipated macro- or micro-economic changes and not improper assumptions about operating costs (because an experienced sponsor should be able to predict operation and maintenance costs fairly accurately). Thus, it is more likely that a prolonged spike in energy or commodity prices or drop in market demand for a service or change in taxes (or royalty rates) will bankrupt a project and not the miscalculation of operation and maintenance costs.

Coverage Ratios

Lenders have designed many ways to try to protect themselves from both market forces and natural catastrophes. As a financial buffer, once a project is complete, lenders will require the SPV to maintain separate cash reserve accounts for items such as debt service and major maintenance. The levels of these reserves will usually be set by the lenders' determination of the project's susceptibility to fluctuations in its revenue and costs. For instance, lenders might require an SPV to retain a reserve (or letter of credit in their favor instead of a cash reserve) equal to the highest amount of quarterly debt service projected to be payable over the life of their loans (which is not the case in a so-called "borrowing-based" lending in which lenders look at a company's "base" of accounts receivables in order to determine what amount of loans they will allow to remain outstanding, or in a collateral-based lending in which lenders look at the market value of a company's inventory [such as oil] in order to determine what amount of loans they will allow to remain outstanding). Project finance lenders typically look at debt service "coverage" ratios to determine how much they will lend and, once construction is complete, permit to remain outstanding. Debt service coverage ratios will gauge the proportion by which the SPV's net cash flow (revenues minus expenses) will exceed its debt service requirements (see Table 4.1).

Debt service coverage ratios can be used prospectively (projecting into the future) or retrospectively (reviewing what has happened in the past). Initially, before loans are made to the SPV, a target debt service coverage ratio will be agreed upon. Once this target ratio is agreed, the amount of loans available will be determined. For example, if a target ratio of 1.3:1 is agreed upon, project revenue and expenses can be forecast and then the amount of debt whose service yields this 1.3:1 ratio can easily be determined and lent.[2] The debt service coverage ratio will be recalculated from time to time (typically every quarter or semi-annually), once the project commences operation in order to check upon the project's financial health. If the ratio falls below a certain minimum threshold (1.1:1 for example) either on "look-back" or "look-forward" basis (that is, for the previous or the upcoming four quarters), the lenders might have the right to declare their loan in default and accelerate its maturity. The rationale for this right to accelerate is that the lenders do not want to be involved in a project that has a fundamental economic problem or is being poorly operated. Thus, loan agreements generally provide that unless loans are prepaid in an amount sufficient to restore the coverage ratios to the required levels (which can be effected by the sponsor's injection of more equity funds into the SPV so that the SPV can make necessary repayments to the lenders), the lenders will have the ability to foreclose on the equity of the SPV or the SPV's assets (because they usually have no recourse to the owners of the SPV) and sell the SPV or its assets to someone who is willing to repay some or all of the loans (or at least is creditworthy enough to assume their repayment).

If the debt service coverage ratio is above a certain level (say 1.4:1), the SPV will typically be entitled to make a cash distribution to its owners. Typically, until the coverage ratio has remained at a satisfactory level for some predetermined number of periods, the SPV's cash will be "trapped" in the SPV as a buffer for lenders in case they need it. Moreover, all of the SPV's funds will be held in accounts that are controlled by its lenders and can be "frozen" by them at any time if a default under the SPV's credit agreement occurs. Funds flow between accounts and out of accounts according to the mechanism of priorities set out in the loan agreement (known as the "waterfall"). The "waterfall" prioritizes how the SPV may spend its money. Typically, funds will be received by the SPV in a "revenue account" and then transferred by the SPV from the "revenue" account into an "operating account" in order to pay operating expenses. Thereafter, remaining funds will usually be disbursed to other accounts

Table 4.1 Sample debt service coverage

Ratio definition from a loan agreement

"Debt Service Coverage Ratio" means, for any period, the ratio of Operating Cash Available for Debt Service for such period to Debt Service for such period.

"Operating Cash Available for Debt Service" means, for any period, Project Operating Revenues minus O&M Costs during such period.

"Debt Service" means, for any period:

(a) all fees due and payable by the borrower during such period to the administrative agent and the banks; plus

(b) term loan interest due and payable by the borrower during such period; plus

(c) scheduled term loan principal payments due and payable by the borrower during such period; plus

(d) net payments made by the borrower during such period pursuant to hedge transactions; and minus

(e) net payments received by the borrower during such period pursuant to hedge transactions.

"Project Operating Revenues" means, without duplication:

(a) all payments received by the borrower under the offtake agreement;

(b) proceeds of any business interruption and liability insurance (to the extent such liability insurance proceeds represent reimbursement of third party claims previously paid by the borrower); and

(c) the investment income on cash and investments.

"O&M Costs" means all maintenance and operation costs relating to the facility and incurred by or on behalf of the borrower in any particular period to which said term is applicable, including:

(a) payments for fuel, additives or chemicals and transportation costs related thereto;

(b) payments in respect of the borrower's franchise, licensing, sales, property, excise, real estate and other similar taxes, and in respect of insurance and consumables;

(c) payments made in connection with the requirements of any permit;

(d) payments under any lease;

(e) payments pursuant to the agreements for the management, operation and maintenance of the facility;

(f) payments for goods or services, including legal and other consulting fees and expenses paid by the borrower in connection with the management, maintenance or operation of the facility, fees paid in connection with obtaining, transferring, maintaining or amending any permits and reasonable general and administrative expenses; and

(g) all costs of settlement of pending or threatened actions, suits, claims and proceedings instituted by non-affiliated third parties against the borrower and all related fines, judgments and other costs (including, without limitation, attorneys' fees) associated therewith and indemnification payments associated therewith; provided that such expenses shall be exclusive in all cases of non-cash charges, including depreciation or obsolescence charges or reserves therefor, amortization of intangibles or other bookkeeping entries of a similar nature, and also exclusive of all interest charges and charges for the payment or amortization of principal of debt of the borrower.

O&M Costs shall not include (i) costs of major maintenance, or (ii) payments for restoration or repair of the facility.

(such as the "debt service" and "major maintenance account") before these funds can finally be transferred to a "distribution account," which can be used at the SPV's discretion to make distributions to the sponsor or for any other purpose. Often a separate agreement called the "account agreement" or "depositary agreement" will govern the "waterfall." The retention of funds in the SPV is necessary because the lenders have lent their money to the SPV itself. Once the funds have been distributed to the owners of the SPV, the owners are under no obligation to return these distributions to the SPV unless it has been agreed otherwise in a contract between the owners and the lenders (often called an "equity support agreement"). The debt service coverage ratios are used to take the financial pulse of a project.

Lenders'"Step-In" Rights

Lenders will almost always negotiate the right to take over day-to-day operating affairs of the SPV in the case of an uncured default by the SPV under its loan agreement (or even a default by the SPV under any of its agreements with its customers or contractors). Typically, lenders request this right even though this right may not eventually be enforceable in many U.S. jurisdictions if the project owner chooses to file for bankruptcy protection (as will be discussed in Chapter 13). Lenders will also have "direct" agreements or "consents to assignment" with each of the SPV's contractual counterparties (such as the EPC contractor and the SPV's customers). These agreements give lenders the opportunity to cure any defaults under the SPV's contracts with these contractual counterparties and will provide that the contractual counterparties take directions from the lenders (or their designee) instead of the SPV if the lenders notify the counterparties that the lenders are exercising their rights to take over the SPV's affairs. This right to cure the SPV's defaults under any of these counterparty contracts is often critical, because without a contract for the facility's completion or sale of the project's output, the project could be worthless.

Typically, in problem cases, lenders will not take over a project themselves but will appoint a manager and/or operator to handle the affairs of the SPV while they decide what course of action is best—that is, to foreclose and sell the SPV's equipment in separate pieces; to sell the SPV as a going concern (operating business); to sell their loans; or to simply write-off their loans on their books (and take any losses they have income to offset) and leave the project in bankruptcy. Lenders must also be careful that their "step-in" rights do not conflict with any other project participant's "step-in" rights. For instance, in the context of projects that are concessions and involve infrastructure that is critical to the public because there is no alternate means to replace the service that the infrastructure provides, it is common for the host government to negotiate the right to seize the infrastructure, finish its construction and operate the infrastructure if the SPV and its owners prove incapable of doing so. In cases of critical infrastructure services, the government will pay very careful attention to the method and progress of the EPC contractor's work, on the theory that the infrastructure is really the government's infrastructure and the SPV is merely being paid a profit to build it and operate the infrastructure for the government without the government being exposed to financial risks such as cost overruns, unforeseen difficulties like accidents, and so on.[3] Thus, a government may be much more proactive in requiring the SPV (and hence the SPV will require the EPC contractor) to deliver to the government any documentation necessary to operate and maintain the infrastructure (such as operating manuals and as-built drawings). The EPC contractor should be required to translate such documentation to the extent required by the government under the concession agreement.

Financing Agreements

Conceptually, all financing arrangements are the same—one party makes money available to another party. Structurally, however, there are many ways to achieve this result.

Bank Financing

Typically, a sponsor that desires to finance part of the capital cost of its project will seek (or engage a financial advisor to seek) a bank that is in the business of originating and syndicating loans to other banks and that will charge an "arranging" fee for this service (a so-called "arranging" bank). Thus, a sponsor may find an arranging bank who will commit, by means of a commitment letter, to lend a substantial portion of the money that the sponsor needs to build its project. The arranging bank then will proceed to negotiate loan documents with the SPV. Once the loan documentation is executed, the loans will be disbursed once the sponsor can satisfy the conditions precedent agreed in the loan documents. While the SPV is in the process of satisfying the conditions precedent in the loan agreement, the arranging bank will find a syndicate of participant banks (and charge them a fee) to "share" in making these loans. Sometimes, the arranging bank will not disburse any loans until it has found participant banks. Often, on smaller projects before loan documentation is signed, the arranging bank has found a group of banks that plan to lend money to the project and this is usually referred to as a "club deal" in the banking industry.

Generally, loan participants will have voting rights when it comes to amending or enforcing the loan agreement with the SPV.[4] Therefore, it is important for the SPV to know what voting thresholds (such as unanimity, simple majority, or some "super-majority" threshold between a simple majority and unanimity) apply to actions to be taken by the lending banks (such as acceleration of the loan or amending the terms of the loan). As a result of this, the SPV may want to restrict which banks can participate in the lending syndicate in order to make sure that only banks with which the sponsors have a good relationship are involved. Since projects are complicated and unforeseen issues can arise easily, it is typical that at some point the SPV will need to request that the lenders waive one or more restrictions contained in the SPV's loan agreement because the SPV cannot meet them. Consequently, a good working relationship with experienced lenders who understand the SPV's business is important. The SPV should always try to negotiate to give the "agent bank" (the bank that acts as the agent for the rest of the banks for the loan—this will usually be one of the banks that arranged the loan, but occasionally arranging banks are not in the business of acting as agents or their fees are expensive and not competitive) as much leeway as possible to make determinations on its own if the SPV cannot meet a loan requirement during construction or operation so that it does not need to submit a vote to the bank syndicate.[5] Arranging banks will usually write a synopsis of the project to be used in connection with their marketing of the loan to other banks. This synopsis is usually referred to as the "information memorandum" or "IM" or "info memo" and usually the SPV will warrant in the credit agreement that the information memorandum (or at least the information furnished to the agent bank in its preparation of the information memorandum) is accurate and not misleading in any material regard.

Construction and Term Loans

Most commercial banks are not willing to lend money on a long-term basis. Typically, they will lend for the construction period and these "construction" loans will then be "converted" to term loans once the project passes its performance tests and its cash reserve accounts are filled (thus "conversion" from a construction loan to a term loan is sometimes called "terming out"). These bank term loans usually mature well before the useful life of the project or its long-term offtake or through-put agreements (and often have interest rates that increase gradually over the term) and thus the SPV will have to "take out" these initial loans with "permanent" financing if it can. (This structure is often referred to as a "mini-perm" because it is not meant to be part of the permanent capital structure and long-term capital structure like a mortgage but contemplates a future refinancing.) Luckily for the SPV, there are investors (such as insurance companies) that do not like to bear construction risk but do find long-term and stable cash returns from operating projects attractive, and will often be prepared to refinance construction or "mini-perm" loans. These lenders focus on their yield on the loans and in fact often require the SPV to pay a "make-whole" premium to them to account for their loss of interest yield (and need to find another suitable investment) if the SPV chooses to prepay the loans.

Public Market

Rather than tap the bank market for funding, the SPV may be able to sell bonds or debentures in the institutional or public markets if the *quantum* of debt it needs to raise is large enough.[6] These capital markets usually can offer the SPV longer loan tenors and more operational flexibility than can a commercial bank financing because covenant restrictions contained in public financing agreements are usually not as onerous as covenant restrictions included in bank financing agreements. A significant drawback of public financings is the highly regimented and regulated public consent solicitation process that is required if an SPV needs to request a waiver of a bond covenant from its bondholders. For these reasons, it is more typical that a public financing will be used to replace ("take-out") a commercial bank financing once construction of a project is complete and the project is operating because, at this point, it is much less likely that any waivers of covenant restrictions will be required from bondholders.

Rather than being contained in a credit agreement, the covenants and other provisions relating to bonds are contained in an indenture. In the United States, an indenture will contain certain provisions that are required by the Trust Indenture Act of 1939 (which was enacted to protect the investing public). If bonds are to be offered to the public in the United States, their issuance must be registered with the Securities Exchange Commission (the SEC) pursuant to a detailed registration statement under the Securities Act of 1933 (33 Act). Part of this registration statement will contain a prospectus describing the sponsor, its project and the bonds. The SPV, the sponsor and underwriter (which markets the bonds for the SPV) will usually all be liable to the bondholders under Rule 10(b)(5) of the Securities Exchange Act of 1934 for any losses that the bondholders incur if the registration statement fails to disclose a material fact to the bondholders (so-called "Rule 10(b)(5) liability").[7] Proper disclosure of risks is an SPV's only defense against liability—prospectuses are documents written by lawyers to be defended in court—not by bankers to market loans.[8]

Infrastructure projects tend to be "sliced up" many different ways for both taxation and financial reasons once they are operating. Most typically (except in the case of very, very large projects), rather than an offering of bonds to the public at large, a select group of institutional investors (such as insurance companies) will be solicited by underwriters to lend money to a project. So long as this "solicitation" meets certain "safe harbor" rules promulgated by the SEC and the bonds are sold to "Qualified Investment Buyers" (QIBs as they are called under the SEC's regulations), no public offering will be deemed to have occurred and, therefore, no registration statement needs to be filed by the SPV with the SEC. While obviating the need for this filing will save the sponsor significant legal costs and time, it is likely that institutional investors will still require much the same information as required in the registration statement. Therefore, the sponsor will still often prepare an investment memorandum even though it will not be filed with the SEC.

Functional Analysis

Once initial financial analysis is in progress, the sponsor must do some technical analysis and preparation. The sponsor must determine, on a conceptual basis, the key attributes of the facility that it desires and how it wants its facility to function. What technology will the facility employ and what type of service will it be designed to provide? In the case of a power plant, for example, will the plant be "baseloaded" and operate around the clock to serve primary electricity demand requirements, or will the plant be a "peaking" station designed to be switched on quickly and efficiently to meet spikes in electricity demand such as those that occur on power grids during peak hours of hot summer days? The sponsor must put together a basic profile of operational demands that will be made on its facility.

Once a profile of the facility's operating demands has been prepared, it will be used by the sponsor's operational experts and consultants to compile a description of what functions the facility must be able to perform and what specifications and requirements (of law, industrial code and local infrastructure systems) the facility must meet. What should emerge from this step is an overview of how the facility will function, commonly referred to as a "functional" or "performance" specification. Its purpose is to specify what features the sponsor wants incorporated into its facility. The functional specification typically does not explain how to engineer or construct the facility. Usually, with the exception of major items, the functional specification specifies equipment by type or function but not vendor or model. The functional specification outlines the sponsor's requirements and will continue to evolve throughout the development process. Eventually, the functional specification will be used to develop the standards that the sponsor will use to determine whether or not the EPC contractor has designed and constructed the facility that the sponsor contemplated. The functional specification will be used to assess whether or not the facility is capable of meeting the sponsor's original profile for the facility's operations.

A functional specification is different from a "design" specification. In a design specification, the sponsor specifies exactly what the sponsor desires by providing highly detailed plans and specifications. In the case of a design specification (typically used for constructing buildings or roads), the contractor must build exactly what was specified by the sponsor's engineer or architect in the specifications and plans. Unfortunately for the sponsor, providing a design specification (as opposed to a functional specification)—even if the design specification is not included in the EPC contract when it is ultimately executed—can often relieve the contractor of liability should the project not perform as contemplated by the sponsor. This shifting of the risk of inadequate or improper performance from the contractor to the sponsor can have serious consequences for the sponsor's budget if expensive reparations must be made, because the sponsor will bear responsibility for their cost. Thus, a sponsor must be careful to avoid carrying out more design work than is necessary to communicate to the EPC contractor what the sponsor expects in terms of the facility's configuration and performance.[9]

Approaching EPC Contractors

With its facility's profile complete and an initial outline of a functional specification, a sponsor can begin to approach EPC contractors to solicit their thoughts and comments in preparation of a price quotation indicating the cost of the facility. At this point the sponsor has several alternatives, each with its own advantages and disadvantages. The right choice will depend upon the sponsor's own constraints in terms of the funds available to be put at risk during this high-risk development phase, requirements of the sponsor or the offtaker regarding the facility's in-service date, and the availability and cost of capital to fund the project.

Formal Bid Solicitations

A formal tender solicitation process (often known as a "request for proposals" or "RFP" or "invitation to tender") can be employed to solicit bids from EPC contractors. This approach is most often employed by governments, agencies, utilities and non-profit institutions whose goal is to foster competition and ensure impartiality in the selection process. This formal RFP approach usually involves several steps. First, a brief description of the project is composed and made available. Next, interested parties are required to furnish a statement of their financial and technical qualifications. From these "pre-qualification" responses, usually a "short-list" of qualified (often referred to as "pre-qualified") bidders is selected. These pre-qualified bidders then receive a detailed "request for proposals" which will contain details of the project, its functional specification and any other relevant requirements regarding the project (such as its required in-service date or edicts governing the use of equipment or construction practices). Pre-qualified bidders will then be given a period in which to prepare technical and commercial responses. Any questions that arise during the bidders' analysis of the project usually must be submitted in writing to the sponsor; the sponsor's written responses (along with the original questions) are generally shared with all bidders in order to prevent any bidder from gaining an advantage.

Often, bidders are required to submit bid bonds with their bids. This bond will guarantee that the bidder will maintain the validity of its proposal during the time required by the sponsor to evaluate the bids and then negotiate the EPC contract with the winning bidder. Some escalation of the price is often permitted for bids that must remain "open" for more than a short time in order to take into account that commodity prices and wages tend to rise. The bond also will guarantee that, if chosen, the bidder will execute the EPC contract according to the terms of the bid that the bidder submitted. "Bonding" is used to prevent bidders from making irresponsible bids and wasting the sponsor's time in evaluating questionable bids that the bidder does not intend to honor but rather has bid as a "bait and switch" type tactic to negotiate its bid with the sponsor. Of course, sponsors must also be wary of unscrupulous EPC contractors who submit low bids in order to be selected but intend (as will be discussed in Chapter 12) to compensate for their low EPC price by seeking changes to the EPC contract once work begins.[10]

In structuring an RFP, the sponsor must decide how it would like bidders to consider its project. Many approaches can be taken. First, the sponsor can simply note all requirements for the facility's performance and solicit pricing terms from EPC contractors based upon the technology and design they will employ. Second, the owner can establish target levels of performance for the facility that bidders must meet for their bids to be considered but give preference to bidders who can exceed these levels. Finally, the sponsor can proceed without specifying performance levels and then use the performance levels contained in the bidders' submissions to test the sponsor's own expectations about plant performance. This approach will inform the sponsor of what technology EPC contractors believe to be commercially viable under the circumstances.

In evaluating bidders' responses, a sponsor has several options. The sponsor can inform bidders of the guidelines or the methodology it will use in evaluating bids, or the sponsor can refrain from giving any indication of how it will rank bids. Governmental agencies, however, are usually bound to use objective and published criteria in awarding projects. Often they are required to disqualify bidders that submit non-conforming bids.[11] In all cases, sponsors must be careful they do not appear to be acting whimsically, because such behavior may deter bidding on their future projects or may simply encourage bids that are not well-prepared or are non-conforming if bidders expect there is little chance that the bidding process will be conducted fairly. So, while bids for infrastructure projects are typically not merely auctions in which the lowest bid is accepted, EPC contractors often take the view that bid requests from sponsors with a history of acting arbitrarily are not a good use of their time and usually put only the most cursory of efforts into their bid preparation in responding to bid requests from capricious sponsors.

Often, commercial and technical portions of a bid will be evaluated by separate financial and technical experts so that the best technical solution is not prejudiced by a high- or low-priced bid. After the evaluation process, a "preferred" or "winning" bidder will usually be selected in an attempt to negotiate an EPC contract based upon the bid. In case the negotiation with the preferred bidder does not result in an executed EPC contract, the second place bidder will be invited to try to agree upon an EPC contract or, as an alternative, the bidding process will be repeated.

The sponsor should not underestimate the tenacity of initially unsuccessful bidders who often are willing to improve the competitiveness of their bid once they learn that they have not submitted the most competitive proposal. The sponsor should not prematurely eliminate what it believes are uncompetitive proposals, because unsuccessful bidders often revise their proposals in the hope that they will garner reconsideration from the sponsor.

While the formal RFP process is designed to elicit the most optimal bids, it is questionable whether this is always successful. Market conditions can influence the efficacy of the formal RFP approach. In a market in which EPC contractors are busy and requirements of submitting a bid are onerous, some (especially the best) EPC contractors may choose not to participate in a formal solicitation, because preparing a bid requires human resources and time. All EPC contractors have business development budgets and internal corporate groups responsible for submitting proposals to project developers. All EPC contractors must have a pipeline of prospects because so many projects never materialize or are delayed. However, when times are good and construction labor and project management professionals are in short supply, submitting proposals along with many other bidders may not be the key to maintaining a profitable EPC business for an EPC contractor. The odds of success are lower in public bids because there are usually more competitors.

No responsible bidder can win every bid. Any bidder that does begin to win every bid it submits is usually headed for (or already in the midst of) a serious financial problem. In fact, sponsors sometimes require the winning bidder to turn over to an escrow agent all documentation the bidder prepared in connection with making its bid in case the EPC contractor later attempts to make a claim against the sponsor alleging that the EPC contractor was not aware of a certain condition or fact. While there is extensive case law that supports the proposition that a contractor will be responsible for assessing the complexity of the work,[12] there is also jurisprudential support for the notion that sponsors must not withhold crucial information from contractors in an RFP. Disclaimers from sponsors that information is supplied for "information purposes only" will not insulate sponsors from liability for failing to be candid with contractors. This is especially true in the case of a "positive assertion" by a sponsor that turns out to be a blatant misrepresentation once the contractor commences work. In these cases, the sponsor will be liable to the contractor for the contractor's costs. Thus, it is advisable for the sponsor to investigate the certainty of any statements that it makes in an RFP.[13]

In a market in which there are many projects and too few responsible EPC contractors, formal RFP solicitations might not receive the attention that their issuing sponsors intend. From the sponsor's perspective, formal RFPs take time and money to prepare and evaluate, a serious drawback for a project that must be implemented quickly.

Sole Source, "Open Book" Negotiations

While there are many ways to accelerate and streamline the formal RFP approach, a sponsor can instead choose the "sole source" approach. In a "sole source" negotiation, the sponsor informally interviews a few EPC contractors who, it believes, can perform the work and then selects one with which to commence negotiations. The sponsor will share the functional specification and all other relevant technical information with this "selected" EPC contractor and the EPC contractor will "open its books" to the sponsor to reveal the estimated cost of each component of the work so that the sponsor can understand all the costs involved in the project (and either acknowledge that they are reasonable or suggest or seek technical alternatives that might be less costly but that will still achieve the sponsor's objectives in terms of the performance and expected maintenance of the facility).

This approach is particularly common in the refining, petrochemical and LNG industry in which a fairly significant amount of so-called "front-end engineering design" or "FEED" is necessary because different technology must be evaluated in order to determine which might be best. Often this work can take six months or more and it is typical for an EPC contractor to be paid for this "front-end" definitional work in the hopes of becoming the EPC contractor. One danger inherent in this approach is that the design engineers might not have a good sense of the "constructability" of their design. It is not uncommon in cases like this for an owner to share a design specification with an erection company that will find many difficulties in being able to carry out the design feasibly or in accordance with the schedule desired by the owner. Thus, there is always an advantage to using the design engineer as the EPC contractor so that the EPC contractor will not later be able to request an equitable adjustment on the basis that the design specification provided by the owner contained errors or issues for successful construction.

Once the sponsor and the EPC contractor agree on the pricing of major components of the work, a contingency amount and a profit component will be added to these costs and a final EPC price will be set (and the "books" will be "closed") so that the EPC contract can be signed. Absent extraordinary circumstances (to be discussed in later chapters), the sponsor has now locked in a price for the project and the EPC contractor will take the risk that its estimates were erroneous. If the actual costs turn out to be less than the estimates, the EPC contractor will have a better profit. If the actual costs exceed the estimates, the EPC contractor's profit will suffer.[14]

The advantage of the sole source approach is that the sponsor has a chance to see the risks and costs involved in all aspects of its project from the EPC contractor's experienced perspective. The sponsor can then identify the areas of concern and try to minimize them. This approach can save time and money and even expose fatal technical or economic flaws long before they would be exposed in the formal RFP approach. In general, EPC contractors, particularly those that are successful, prefer the open book approach because it is likely to result in a signed EPC contract in a much shorter period than would a formal RFP. This accelerated selection and negotiation schedule usually benefits the sponsor as well. Development is a long and risky process. The longer it lasts, the more likely it is that problems will arise and costs will rise concomitantly.

But the sole source approach is not without its shortcomings. First, under the formal RFP approach the sponsor often has the opportunity to review the suggestions and solutions of several different EPC firms, each of which may address the sponsor's requirements differently and with varying degrees of success. In that case the sponsor will be able to isolate the best aspects of each bid and incorporate those ideas or methods into the winning bidder's EPC contract. This is generally not possible under the sole source approach. Most importantly, however, the sole source approach tends to remove from the bidding process what is perhaps one of the strongest forces known to mankind—competition. Without the immediate threat of losing the EPC contract award on the basis of a price that is too high, the sponsor is essentially relegated to using a hand chisel and mallet to chip away at the EPC contractor's price instead of the jack-hammer that the sponsor usually needs to blast through to the rock-bottom capital cost of a project.

Focusing the EPC Contractor's Attention

No matter which bidding approach is employed, it is important for the sponsor to realize that an EPC contractor must compile and process information from many different parts of its organization in order to submit a responsive proposal. This initial bid stage may be the only time that all the individuals involved in the EPC contractor's different disciplines (such as financial, taxation, engineering and procurement) will focus their attention on the sponsor's project at the same time in a coordinated effort. The sponsor can take advantage of the fact that the EPC contractor's entire team is informed and eager to resolve any issues (often at the EPC contractor's own expense) by identifying these issues and insisting that they be settled before negotiations on the EPC contract can commence—thus capitalizing on the mental leverage that the EPC contractor created through its proposal group. If the issues are left to be resolved at a later point in negotiations, the EPC contractor's proposal team may lose some of its cohesiveness and familiarity with the project as they move on to other proposals. Solutions may not be as forthcoming with the same rapidity or ingenuity as was initially possible, an unfortunate result for the sponsor (which never has time on its side).

Over time, it becomes more and more difficult for a sponsor to disengage from an EPC contractor (even if the EPC contract has not been signed) without incurring unnecessary expense and delay. EPC contractors are aware of this reality. The sponsor, therefore, should beware of leaving items (such as milestone payment schedules, liquidated damage amounts and liability limitations) for later stages of negotiations. For several reasons, these issues should probably be resolved before negotiations commence or continue. First, doing so avoids costly surprises. Second, if the sponsor is still in the process of eliminating potential EPC contractors, different contractors may resolve open issues differently and the owner will not have the benefit of comparison points if it eliminates bidders too quickly. Thus, the sponsor may be unnecessarily losing viable solutions by eliminating bidders.

While some issues cannot be addressed at the bidding stage, the sponsor is well-advised to formulate a list of critical issues for its project and specify for bidders which of these must be addressed by them in order for their bid to receive any consideration. This practice will help EPC contractors focus on the sponsor's true concerns and allow the sponsor to discern which bidders are likely to be best suited to carry out its project.

Communicating with the EPC Contractor and the Attorney/Client Privilege

A brief note about the attorney/client privilege may be useful. This privilege was developed to protect the confidentiality of legal advice given by attorneys to their clients and foster an honest and candid relationship between attorneys and their clients so that attorneys can act appropriately. In civil cases, facts are not protected by this privilege. For instance, if a communication between an owner and its attorneys lists several legal alternatives for pursuing a warranty action against the EPC contractor based upon the fact that the owner's expert has discovered that the EPC contractor has "driven" pile lengths of 28 feet instead of 35 feet as was required and as a result of this failure the facility is "settling" unevenly, the EPC contractor might not be entitled to receive a copy of this communication but the EPC contractor could probably obtain the "facts" that the piling lengths are 28 feet. Facts are usually not privileged. Furthermore, "internal" communications between a party's employees are also not privileged and can be "discoverable" (that is, they must be provided) by an opposing party who may (and typically does) ask to see them. Should an employer wish to try to protect an internal communication between its employees, it is best to send the communication to its legal counsel to forward a copy of the communication to the other employees that should see it. This is more likely to protect the communication from disclosure during discovery requests. Following this "routing" procedure with all internal communications is not a good idea, however, because a judge may decide that a party was abusing the attorney/client privilege to the detriment of its contractual counterparty and hold that all such communications routed through the party's attorneys are discoverable by the counterparty (except those that the court decides really do contain legal advice). EPC contractors should be particularly wary in cases of delay claims. During the negotiation stage of the EPC contract, when the EPC contractor was trying to "bag" the "job," the EPC contractor may very well have had to propose an aggressive project schedule to the owner. With the proliferation of electronic mail instead of oral communication, it may be possible that communications between the EPC contractor's proposal manager and the EPC contractor's construction professionals have preserved statements like "Boss, we can never do the project by that date" for posterity. More important, for a judge or jury, a casual statement like this can be transformed by a zealous plaintiff's lawyer and exploited in a statement such as "prophesies of knowledgeable professionals which were ignored in bad faith by avaricious managers acting with willful disregard for the dire financial consequences that the owner might suffer in the case of a delay in the completion of its project." As will be seen in Chapter 20, this alleged "bad faith" of the EPC contractor can be used as one tactic to try to invalidate the limitations on the liquidated damages negotiated by the EPC contractor in the EPC contract.

Notes

1 Availability payments are typical in passenger railroad and toll road concessions in which sponsors have run into serious mistakes in trying to predict ridership and traffic and therefore to attract sponsors; governments offer fixed payments based on availability and quality of service so sponsors and their lenders are not exposed to the risk of traffic or ridership patterns.

2 Of course, debt service will depend upon the interest rate and the loan's amortization schedule (that is, mortgage style of equal payment vs. proportional payment of principal). Typically, while lenders will lend at a "floating" interest rate that fluctuates with the lender's actual or nominal cost of funds (such as LIBOR), lenders will usually require the SPV to hedge its exposure to this fluctuation by paying a charge to "swap" this floating rate obligation for a fixed rate obligation with a counterparty that agrees to take the risk of the floating interest rate. Depending upon the projected financial robustness of the sponsor's project, lenders usually require that an SPV hedge between 50 and 75 percent of its interest rate exposure.

3 In fact, under generally accepted accounting principles (GAAP) in the United States, it is recognized that an offtaker that has contracted with a producer to consume the bulk of the output from a producer's facility on essentially an exclusive basis for the facility's useful life may have, in essence, entered into a lease for the facility with the producer and therefore, if this turns out to be the case when the facts and circumstances of the particular arrangement are scrutinized by the accountants, may have to account for the facility on its own books and records as if the offtaker (and not the developer) owns the facility and has borrowed money to finance the facility (see Financial Accounting Standard Board (FASB) Statement of Financial Accounting Standards No. 13 "Accounting for Leases" and Emerging Issues Task Force Release 01-8). This consolidation of the debt of the developer onto the balance sheet of the offtaker has the potential to discourage offtakers from entering into long-term offtake agreements once they learn of the consequences that these arrangements can have on their own balance sheets. Thus, it is advisable for a sponsor to raise this "accounting" issue early in the process with an offtaker so that an appropriate understanding and resolution can be reached with the offtaker's accountants and financial managers. Similar consequences may arise under International Financial Reporting Standards (IFRS).

4 See *In re Enron Corp.*, 292 B.R. 752 (Bankr. S.D.N.Y. 2003), in which a bank participating in a letter of credit attempted to dispute the issuing bank's payment to an owner under the letter of credit that had been posted in its favor by an EPC contractor.

5 Generally, a single bank will be appointed as an "agent bank" to act as the agent for the other lending banks in the administration of the loan to the SPV. Sometimes other banks will be appointed for other functions such as the "collateral agent bank" to manage the banks' collateral posted by the borrower for the loans or the "technical agent" to review engineering and operational questions.

6 The terms "bond" and "debenture" are usually used interchangeably but "debentures" are generally notes for which no collateral has been pledged to secure their repayment.

7 The underwriter may be able to avoid liability if it carried out a diligent investigation of the project and still failed to learn of the risk that subsequently materialized.

8 Generally, loans are not considered to be securities subject to U.S. federal securities laws.

9 See *Trataros Construction, Inc.*, General Services Board of Contract Appeals, No. 14875, 2001-1 BCA ¶ 31,306 (2001), in which it was decided that a contractor was not responsible to determine the adequacy of the owner's design work because the owner had provided a design (and not a performance) specification.

10 See *Sunhouse Construction, Inc. v. Amurest Surety Insurance Company*, 841 So. 2d 496 (Fla. Dist. Ct. App. 2003), in which a court denied claims of a contractor for "extra" work because the work was to be included under the contract as written.

11 See *Halter Marine, Inc. v. United States and Marinette Marine Corp.*, 56 Fed. Cl. 144 (2003), noting that "a disappointed bidder has the burden of demonstrating the arbitrary and capricious nature of the agency decision by a preponderance of the evidence." Also see *CSE Constr. Co., Inc. v. United States*, 58 Fed. Cl. 230 (2003), awarding bid preparation costs to a bidder because the defendant acted in an arbitrary and capricious manner.

12 See *Superintendent and Trustees of Public School of City of Trenton v. Bennett*, 27 N.J.L. 513 (Sup. Ct. 1859). "No rule of law is more firmly established by a long train of decisions than this, that where a party, by his own contract, creates a duty or charge upon himself, he is bound to make it good if he may, notwithstanding any accident or inevitable necessity, because he might have provided against it by his contract."

13 See *Hollenback v. United States*, 233 U.S. 165 (1914), in which it was held that the United States was liable to a contractor for making a misstatement about the constitution of a dam. Also see *Golomore Associates v. N.J. State Highway Auth.*, 173 N.J. Super. 55 (App. Div. 1980), holding that a contractor was entitled to recover damages in a case that the amount of fill that the owner had indicated was available from a site was incorrect. But also see *Sasso Contracting Co. v. State*, 173 N.J. Super. 486, 489 (App. Div. 1980), holding that a contractor should have done its own investigation and could not recover damages under a contract that clearly stated that conditions were noted as a "guide for design" and "not warranted to be accurate." And see *Ell-Dorer Contracting Co. v. State*, 197 N.J. Super. 175, 183 (App. Div. 1984), in which a contractor was denied recovery for damages for work not performed because the contractor needed to undertake less excavation than the owner had estimated and the contractor desired more work.

14 See *Lord v. Mun. Utilities Auth. of Lower Twp., Cape May County*, 128 N.J. Super. 43, 46 (Ch. Div. 1974). "Basically such a [lump sum] bid is a wager because of the various unpredictable contingencies which the general contractor must weigh against the priority of submitting the lowest bid in order to obtain the contract at a reasonable profit."

Chapter 5

Full-Scale Development

The Development Timeline

Whether the request for proposal or sole source negotiation approach is used, the subsequent development schedule is essentially the same. Once an EPC contractor is selected, time must be spent negotiating the EPC contract and finalizing the functional specification for the facility. This functional specification eventually will be attached to the EPC contract as an exhibit to prevent confusion about the specific facility characteristics that the sponsor desires. See Figure 5.1.

The development process is easily separated into four distinct phases:

1. planning and feasibility;
2. contractual negotiation and permitting;
3. engineering and construction; and
4. operation and maintenance.

During each phase, especially the first two, it is of paramount importance to attempt to minimize expenditures (so long as they can be postponed without jeopardizing the target in-service date for the project). The risk of failure during the first two periods is very high. For that reason, a sponsor's investment of time and money often is never recouped. Not until contracts for the project are executed and permits are obtained can the sponsor reasonably expect to be able to build the project and earn the opportunity to try to recoup its investment at a profit. In a project as complicated as a power station, LNG terminal, railroad, port, mine or refinery, it is not commercially practical to carry out all tasks *seriatim* (meaning consecutively). Therefore, the sponsor must determine which activities are on the critical path to completion and carry out these activities in parallel if necessary. As should be clear, infrastructure development is not for "would-be" developers with a dollar and a dream. It is for sponsors with millions of dollars and good plans.

The Environmental Impact Study

Once the sponsor is reasonably confident it has an economically feasible prospect for a project, it will commission a study to assess environmental consequences that the project might create and address how the project will comply with the requirements of environmental regulation. If any negative impacts are revealed, further design

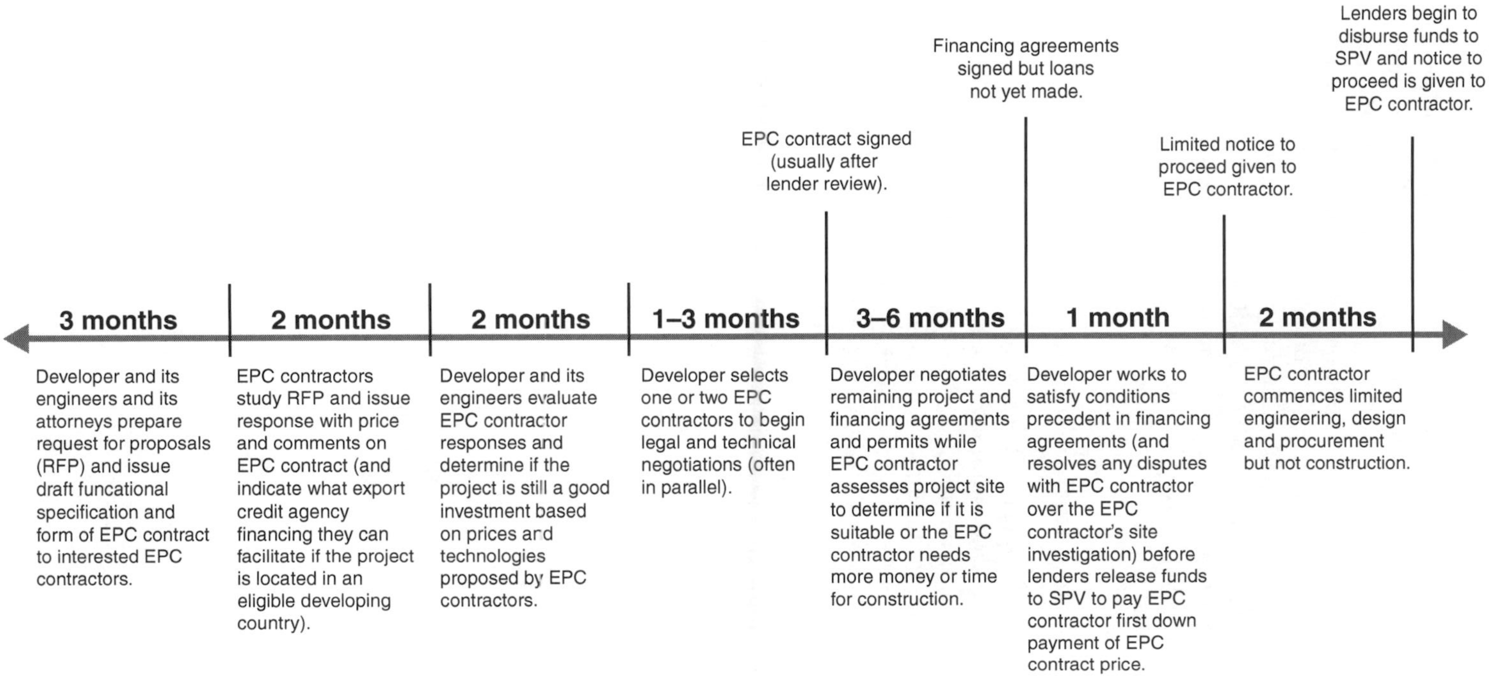

3 months

Developer and its engineers and its attorneys prepare request for proposals (RFP) and issue draft funcational specification and form of EPC contract to interested EPC contractors.

2 months

EPC contractors study RFP and issue response with price and comments on EPC contract (and indicate what export credit agency financing they can facilitate if the project is located in an eligible developing country).

2 months

Developer and its engineers evaluate EPC contractor responses and determine if the project is still a good investment based on prices and technologies proposed by EPC contractors.

1–3 months

Developer selects one or two EPC contractors to begin legal and technical negotiations (often in parallel).

EPC contract signed (usually after lender review).

3–6 months

Developer negotiates remaining project and financing agreements and permits while EPC contractor assesses project site to determine if it is suitable or the EPC contractor needs more money or time for construction.

Financing agreements signed but loans not yet made.

1 month

Developer works to satisfy conditions precedent in financing agreements (and resolves any disputes with EPC contractor over the EPC contractor's site investigation) before lenders release funds to SPV to pay EPC contractor first down payment of EPC contract price.

Limited notice to proceed given to EPC contractor.

2 months

EPC contractor commences limited engineering, design and procurement but not construction.

Lenders begin to disburse funds to SPV and notice to proceed is given to EPC contractor.

Figure 5.1 Example development life cycle.

work can be conducted to evaluate whether or not the deleterious circumstances can be feasibly mitigated or the project must be abandoned.

Many parties aside from the sponsor and EPC contractor will be privy to this environmental study's contents such as public agencies and non-governmental organizations (NGOs). Consequently, it is essential for the sponsor to ensure that the report is well-researched and prepared by a reputable expert who will be able to justify and defend its findings eloquently and lucidly before different audiences and interest groups.[1] This has become particularly important as public sentiment seems to make siting infrastructure projects more and more difficult. Expressions such as "NIMBY" ("not in my backyard") have now been replaced in the infrastructure development lexicon with acronyms like "BANANA" ("build absolutely nothing anywhere near anything") and "NOPE" ("not on planet earth").

The Contractual Negotiation and Permitting Phase

Once a project has been determined by a sponsor to be feasible and a potential EPC contractor has been identified, contractual negotiations can begin. At this point, the sponsor usually faces a difficult decision: deciding whether to embark upon the most costly and risky part of the development process—permitting and siting. Although the sponsor can do its best to manage and orchestrate this process, generally the sponsor will be at the scheduling mercy of governmental agencies, public hearings and possibly environmental activists. The sponsor frequently will have to retain legions of lawyers, public relations experts and environmental consultants to guide the project safely through regulations and protests to receive its permits (which are the lottery tickets for the developers' paradise). To begin the permitting process, the sponsor must know what type of equipment the EPC contractor will be installing and such equipment's emissions profile. Based upon the emissions information, applications to permitting agencies can be made. As a result, the sponsor and EPC contractor often must begin to work together even before the EPC contract is signed. Generally, EPC contractors will work with sponsors on an informal basis at this permitting stage in the hope that their unpaid efforts will facilitate a prompt execution of the EPC contract with them by the sponsor. However, in cases that require significant engineering time and detailed submittals to permitting agencies, it is not uncommon for the EPC contractor and the sponsor to enter into a preliminary or limited services agreement of the type set forth in Volume II, Part B. It is important that the sponsor, if it enters into an EPC contract with the EPC contractor, ensures that the EPC contract encompasses these preliminary services under its warranty provisions and that the preliminary services contract itself terminates at the time the EPC contract is executed because the two contracts could contain conflicting provisions (especially in relation to warranties and the liability of the EPC contractor to remedy warranty items).

To provide these preliminary services, the EPC contractor must receive input from several other project participants. In particular, in a power plant, for example, the EPC contractor must understand the technical requirements and limitations of the power customer, the party that will receive and transmit the electricity. The EPC contractor must design the facility to handle the fuel and other inputs involved, such as the fuel's heating value, chemical composition and physical characteristics, as well as the requirements of the fuel transporter involved, usually a railroad, pipeline, marine shipping or trucking company. Finally, the facility's source and consistency of raw or "gray" water must be identified together with any limitations on its subsequent disposal. Thus, it will be necessary for the sponsor to have had substantial discussions (and perhaps even negotiations) with all of these other project participants.

Coordinating dealings and negotiations between all these unrelated entities often places a great strain on the sponsor in terms of human and financial resources. Often the sponsor will feel the mounting pressure of significant development capital at risk without any executed agreements or issued permits to allay such risk. In any event, before construction actually begins, all projects are simply a file cabinet of executed contracts and approved permit applications. Until each necessary agreement and permit is obtained, a project is nothing but a useless pile of paper probably worth little more than its recycling value and the real estate, if any, that has been purchased or leased.

Unfortunately, to complicate matters and put more pressure on the sponsor, unless the sponsor plans on using its own money to pay for the construction of the project, the sponsor must be in frequent communication with potential lenders. As was discussed in the last chapter, to facilitate this communication the sponsor often will put together what is often referred to as an "offering memorandum," "preliminary investment memorandum," "private placement memorandum" or "offering circular," which describes the project, its purpose, its participants and its prospective profitability. However, until the EPC contract is executed and a price tag is placed on building the project, the sponsor's financial projections will not be viewed as credible by potential lenders. If potential lenders indicate interest in the project after considering many factors, such as the reputation and credit standing of the participants and the political, economic and technical state of the market in which the project will be located, it is probably in the sponsor's best interest to execute an EPC contract as soon as possible in order to "lock in" a price for the facility so the sponsor can proceed to solicit funds from potential lenders. Why? Because financial markets change quickly and lenders can quickly withdraw commitments (which are often by their terms revocable if material adverse changes in market conditions occur) and leave the sponsor with no way to pay for the facility's construction. The decision to conclude negotiations with the EPC contractor and execute an EPC contract is the most critical decision that will ever be made concerning the project. If the sponsor does not execute the EPC contract, theoretically (but, of course, depending upon market conditions) the sponsor can continue to negotiate with the EPC contractor to try to reduce the EPC price. Any decrease in the EPC price that the sponsor is able to secure should cause an increase in the sponsor's profits. As leverage to reduce the EPC price, the sponsor can usually invoke the name of, or simply negotiate with, the second-place EPC bidder. The EPC contractor will sometimes call the sponsor's bluff and hold firm on its price knowing that the sponsor does not have the time or resources to begin a second round of negotiations with another EPC bidder, especially if the permitting process is underway and the sponsor itself is competing with another sponsor to implement the project for a "host" utility or government.

If the sponsor executes the EPC contract, it will have, in effect, set the profitability or "unprofitability" of the project forever because once the EPC contract price is set, assuming the project will sell its output or capability for use (such as for electricity or desalinated water, or LNG re-gasification capacity) to one buyer under a long-term, fixed price arrangement addressing fuel and feedstock escalation (or in the alternative sell into the smelter or downstream market [such as precious metals or refined petroleum] and enter into longer-term hedges, either "financial" or "exchange for physical," with financial institutions or on an exchange), the only variables left to manipulate are the amount, term and interest rate of the loans that will finance part of the capital cost to build the project. As was discussed in the last chapter, the amount of loans (as a percentage of capital cost) and the spread above the reference interest rate usually are not highly negotiable with banks. Either they lend to the project or they do not. The sponsor may have some negotiating leverage but rarely enough leverage to alter the internal rate of return on its investment significantly.

Unfortunately for the sponsor, even if it executes the EPC contract and locks in a price, the prospects for financing remain at risk because the potential lenders will want their engineering consultants, insurance advisors and lawyers to review the EPC contract to see how risks have been allocated between the sponsor and the EPC contractor. Lenders usually put more money at risk than the sponsor. It is not uncommon to hear lenders' lawyers remark that the hallmark of project finance is the principle that a project's contractual arrangements must place the key risks on the party most able to bear them in order for lenders to agree to finance a project. Arguably, however, in the final analysis and once lenders' concerns are addressed in the loan arrangements, the risks seem really to reside with the party that is least able to bear them—the sponsor. If the EPC contract has not addressed relevant risks to the lender's satisfaction, the lenders will require amendments to the EPC contract as a condition to making their loans. Often these conditions can require extensive and expensive changes to the EPC contract or even reduce the overall amount of loans the lenders will make to finance the project if the EPC contractor will not agree to them. Lenders may even require the sponsor to contribute more of its own funds to reduce the risks further. EPC contracts generally include a provision requiring EPC contractors to use their good faith efforts to accommodate "customary and reasonable" requests from potential lenders to modify the terms of the EPC contract. Obviously, this is essentially meaningless because once the EPC contract is executed, the EPC contractor is under no legal obligation to amend or modify the EPC contract unless the EPC contract expressly provides so. On the other hand, since sponsors usually put a clause in the EPC contract that provides that they may cancel the EPC contract without liability if they cannot raise financing, there may be no project at all if the EPC contractor does not make these accommodations and thus EPC contractors are often cooperative when lenders make these requests, but will usually pass on the added cost of these requests to the sponsor in the form of an increase of the EPC contract price.

To make matters worse for the harried sponsor, permits may ultimately be issued with more restrictions or more stringent standards than the sponsor expected when permit submissions were filed. In other cases, a sponsor may decide to settle any objections to one or more of its permit applications by agreeing with the objectors to reduce the emissions levels the sponsor initially requested in its permit application. If these submissions formed the basis for the EPC contract price, the EPC contractor will often require an increase in the EPC price to take into account additional costs involved in meeting more onerous standards.

Either of the above cases will precipitate a fall in the sponsor's internal rate of return on the project and perhaps jeopardize the project's ultimate financial viability. Thus, while speedy negotiations of the EPC contract are desirable from both the sponsor's and EPC contractor's points of view, an EPC contract signed in haste can lead to problems when the project reaches its financing and construction stages. In fact, it is important that the EPC contract specifically note that "time is of the essence." This statement will serve to inform the EPC contractor that it is the sponsor's intention that the EPC contractor's ability to carry out the work in a prompt manner is central to the sponsor's decision to retain the EPC contractor. This construct of time being "of the essence" has developed in case law over the centuries. Courts generally hold that time is not of the essence in the performance of a contract unless the parties specify that it is.[2] If the sponsor has noted that "time is of the essence," it must make sure that it does not act inconsistently with this assertion or else a court (particularly true under the laws of England, under which time can go "at large")[3] may determine that time was really not of the essence to the sponsor and possibly disregard other deadlines contained in the EPC contract. Thus, any correspondence from the sponsor to the EPC contractor after the EPC contract has been signed should always note that "time is of the essence."

Notice to Proceed

The sponsor usually does not have the financing or the permits to build the project at the time the EPC contract is executed. Although the EPC contract is a legally binding obligation of both parties once it has been signed, the EPC contractor usually agrees not to begin any of the work called for in the EPC contract until it receives a written notice to proceed with the work from the sponsor. This ability to delay the commencement of work gives the sponsor time to complete its financing arrangements, receive its permits and await any other conditions that need to be addressed or resolved before authorizing the start of detailed engineering and construction. Once a sponsor gives the EPC contractor the "notice to proceed," the sponsor undergoes a metamorphosis from sponsor (or developer) to owner. Once the owner's (and lenders') money is "on the ground," it is very costly to abandon a project.

In essence, by employing the "notice to proceed" construct, the sponsor is able to turn the EPC contract into a unilateral option agreement. This is true because generally there is no obligation placed upon the sponsor in the EPC contract to give the notice to proceed to the EPC contractor (and it is a good idea that the EPC contract make clear that the sponsor is never obligated to make a payment or start the work). In recognition of this, EPC contractors often limit this notice to proceed "option" in two respects. First, as will be discussed in Chapter 21, the EPC contract usually contains a period during which the sponsor can give notice to proceed. Once this period expires, if notice to proceed has not been given, the sponsor is no longer permitted to give a notice to proceed and must renegotiate the EPC contract (or, ideally, just the contract price if the sponsor has negotiated this provision properly in the EPC contract) if the sponsor still wants the EPC contractor to perform the work. Second, the EPC contract usually includes a provision that the EPC contract or price will escalate pursuant to an agreed-upon escalator if notice to proceed is given after a certain date. This protects the EPC contractor from increases in costs as time passes.

This protection is often crucial for EPC contractors (especially in the case of projects like an LNG terminal or undersea cable, both of which use large amounts of commodities [such as steel or copper] whose prices fluctuate daily) because it is unlikely that they have executed all subcontracts with all of their subcontractors at the time they execute the EPC contract. Therefore, the EPC contractor is at risk that the price quotes it has obtained from its subcontractors and vendors will increase above those received before it executes the EPC contract.

The notice to proceed provision contains another serious risk for EPC contractors that most EPC contracts fail to address. Often, there is no express provision of the EPC contract that obligates the owner to use the EPC contractor to carry out the work. Conceivably, the owner could refrain from giving notice to proceed to the EPC contractor and then use the EPC contract price as a bargaining tool to negotiate another contract with another EPC contractor. While a judge or arbitrator might eventually reach the conclusion that the owner's relationship with the original EPC contractor was meant to be exclusive regarding engineering, procurement and construction of the project and that the owner is acting in bad faith by using another EPC contractor, this might be a very costly battle for the original EPC contractor to fight. This clash could be avoided entirely with a few words in the EPC contract providing that an exclusive relationship regarding engineering, procurement and construction of the project is intended to have been created between the parties regarding the work, and the owner may not employ another EPC contractor to build the project for some mutually agreed upon standstill period after the EPC contract is executed if full notice to proceed is not given. A good solution for this issue is usually to allow the sponsor to buy its way out of the EPC contract for a fixed and often nominal price to cover the EPC contractor's proposal costs price (sometimes called a "break" fee in this context).

Sometimes the parties will agree upon the terms of the EPC contract but choose not to sign it until the sponsor is ready to give notice to proceed to the EPC contractor. This practice is often not in the sponsor's interest because the EPC contractor can still attempt (often successfully) to increase the EPC contract price or change its terms as time goes by since it is not signed.

Limited Notice to Proceed

The obligatory haggling with diffident financiers over lending terms and exchange of data with persnickety regulatory agencies about permit requirements often create difficulties for a sponsor. This is because the sponsor has often had to promise the future purchaser or user of the facility's output or capacity that the facility will be in service by a certain date. The penalty for not placing the facility in service by this "guaranteed" date can be severe—daily liquidated damages payable by the sponsor to the offtaker or user or even termination of the customer's offtake or use agreement itself. Even if the sponsor has not contractually committed the output or capacity to a buyer or user, the sponsor is often trying to have the facility in service as soon as possible so it can begin to recoup its investment by selling into the market. (For instance, in the case of a power plant or gasoline refinery, perhaps it is best for it to enter into service in a season [perhaps summer or winter or the non-rainy season] when electricity prices are historically high, or in the driving/vacation season or—in the case of an LNG re-gasification terminal—it is best to enter service in the winter when natural gas prices are historically high.)

In the face of all these threats to potential revenue, the sponsor often must put more capital at risk to ensure that tasks or equipment that have long lead times for their preparation, execution or delivery will be completed as necessary and will not delay the facility from entering service by its target in-service date. In the case of power plants, items such as the design and fabrication of heat recovery steam generator systems and high voltage transformers are notorious in this respect as similarly are large castings for crushers and mills in the mining sector. Also, some items, such as large-scale gas turbines, might at the time have been in such short supply that the sponsor purchased them before the particular project in which they are to be installed was ever contemplated. To deal with such circumstances, the sponsor sometimes gives what is commonly known as a "limited notice to proceed" to its EPC contractor whether or not the EPC contract has actually been signed. Usually, this more limited notice takes the form of a notice to proceed with some specific portion of the work, such as ordering a transformer. Typically, an agreement, often referred to as a "Preliminary Services Agreement" or "LNTP Agreement," is executed between the sponsor and the EPC contractor that provides that the sponsor will pay the EPC contractor to commence all work necessary to hold the sponsor's desired in-service date for the project, even though the EPC contract may be weeks or months away from being signed by the parties (Part B of Volume II contains a form of LNTP Agreement). Or, if the EPC contract has been signed, "full" notice to proceed may still be far off (perhaps because the sponsor does not yet have all necessary permits). Thus, while site construction is expensive and prohibited until permits are obtained, engineering and procurement activities (and often site preparation and sometimes foundations) can be conducted before either permits or possession of the site are secured by the sponsor.

From the sponsor's point of view, it is crucial that the EPC contractor acknowledge and represent to the sponsor in the LNTP Agreement that the EPC contractor can meet the sponsor's desired in-service date for the project if the EPC contractor is allowed to begin the work specified in the LNTP Agreement. As noted above, since this LNTP Agreement will be abbreviated and will not contain the extensive protection (in terms of levels of performance and warranties) that the sponsor will negotiate in the EPC contract, it is important that the EPC contract expressly terminate any LNTP Agreement and provide that any "preliminary" work performed under the LNTP Agreement will be considered "work" for all purposes under the EPC contract and be subject to all of the stringent requirements to which work is subject under the EPC contract.

Under an LNTP Agreement an EPC contractor commonly will limit its liability concerning "preliminary" work to the value of payments the EPC contractor receives from the sponsor for performing such preliminary work. There are two reasons for this. First, the EPC contractor is not in business to do preliminary engineering and procurement work and does not stand to make a significant profit on such work. Truly, the EPC contractor is performing this work, in which the EPC contractor does have the potential to make significant profit, as an accommodation to the sponsor in anticipation of being awarded the EPC contract. Second, from a practical point of view, if the sponsor and the EPC contractor do not ultimately reach agreement on the contract, the EPC contractor does not want to be liable for errors or misinterpretations of this work if another contractor is ultimately retained by the sponsor and the new EPC contractor uses the preliminary work in construction of the facility and a design or "constructability" problem surfaces at some later date.

The LNTP Agreement should contain a provision that allows the sponsor to terminate the LNTP Agreement at any time and pay for any work done by the EPC contractor on a time-and-materials basis according to a termination schedule based upon the extent to which the engineering work has been completed by the EPC contractor and also upon payment by the sponsor of any costs that the EPC contractor incurs in terminating any equipment purchase orders that it may have executed. This type of provision is necessary because a full notice to proceed may never be given, perhaps because a permit is never issued or financing turns out to be prohibitively expensive.

Another case in which a limited notice to proceed is typically given relates to a situation in which the EPC contract has been executed but there is a specific part of the project that the sponsor cannot or does not want to construct yet. For instance, a permit relating to "wetlands" may not yet have been obtained by the sponsor. As a result the EPC contractor will be directed not to conduct any site work concerning the wetlands area of the project. In such an event, from a legal point of view it is better for the sponsor to give the EPC contractor a notice to proceed with all work but exclude certain work (for example, excluding work relating to the construction of the water intake facility) rather than direct the EPC contractor to perform only a limited portion of the work (for example, directing the contractor to construct only the "power island"). This is advisable because the sponsor will not know all of the work involved in designing and building the project. Consequently, if the sponsor gave a direction to the EPC contractor about what work to perform instead of a direction about what work the EPC contractor should not perform, it is possible that a crucial task could be ignored by the EPC contractor because it did not receive specific instructions to perform a specific task from the sponsor. This could lead to delay and a costly dispute that would likely be resolved in the EPC contractor's favor.

While the EPC contractor is operating under a "limited notice to proceed" agreement, the sponsor is forced to bear all the project risks because the risk-shifting provisions of the EPC contract have not yet become effective. Usually, at this point, lenders have not yet lent any money (because they only lend once all permits have been obtained), so lender's capital is not at risk either. Frequently, this is the riskiest financial point for the sponsor, who now might need to decide whether or not to cancel the project entirely because its viability is questionable (perhaps as a result of indeterminable permit delays) or to bring the permitting and financing processes to consummation so that a full notice to proceed can be issued.

Notes

1 See *Center for Biological Diversity v. Federal Highway Administration*, 290 F. Supp. 2d 1175 (S.D. Cal. 2003), denying complaints against highway developers and governmental agencies under the National Environmental Policy Act of 1969, the Department of Transportation Act and the Endangered Species Act of 1973, in connection with a highway in San Diego that they alleged would have a deleterious effect on the Otay Mesa Mint plant and Quino Checkerspot butterfly.

2 See *Holt v. United Security Life Ins. Co.*, 76 N.J.L. 585 (N.J. 1908), in which it was held that late completion of an apartment building did not entitle the owner to terminate the contract because the contract did not make time of the essence. Also see *Koolvent Aluminum Awning Co. of N.J. v. Sperling*, 16 N.J. Super. 444 (App. Div. 1951), holding that a purchaser was entitled to recover its deposit from a seller that delivered goods late because the contract had made time of the essence. But also see *Schenectady Steel Co. v. Bruno Trimpolo Co.*, 350 N.Y.S.2d 920 (3d Dept. 1974), holding that a party's failure to take prompt action to enforce its rights when a contract does note that "time is of the essence" can operate as a waiver of the party's rights under the "time is of the essence" clause.

3 See *Multiplex Constructions (UK) Ltd v. Honeywell Control Systems Ltd* [2007] EWHC 447 (TCC).

Chapter 6

Philosophy of the EPC Contract

The fundamental principle of EPC contracting is to make the EPC contractor responsible for all work associated with the design, procurement, erection and testing of the facility. There will be items of work that the sponsor and others must perform (perhaps the provision of fuel, water and interconnections to local utilities) but the overarching premise of EPC contracting is that the EPC contractor has been engaged to turn the owner's functional specification into an operating facility within the time allotted. In fact, as will be discussed in later chapters, it is crucial that, other than monitoring the EPC contractor's progress, the sponsor should not become too involved in the EPC contractor's work; otherwise, the legal protections that the sponsor has negotiated for itself in the EPC contract can be jeopardized.

Intent

In order to construct an impregnable EPC contract, it is necessary to go back to the most basic tenet of contract law—the intent of the parties. This is the foundation upon which all contracts are built in common law jurisdictions such as the United States and United Kingdom. Unfortunately, far too little time and effort is dedicated to this area of contractual groundwork. By law, in the United States, no contract exists until there is a "meeting of the minds" of the parties to the contract. It is not necessary that the parties have agreed upon all the terms to be contained in a contract for a court to find that a contract has been entered into by the parties. There need only be agreement upon the material terms of the contract. The rest of the terms can be "filled in" in any number of ways, including the course of dealing between the parties and common practices of the trade in which the parties are engaged. In common law jurisdictions, a binding contract is generally formed when one party has incontrovertibly accepted the offer of another party. Except for some limited cases such as contracts for the sale of real estate, contracts need not be in writing in jurisdictions with civil codes.[1]

Quasi-Contract

As will be discussed in Chapter 8 and Chapter 12, in common law jurisdictions a party's unilateral mistake in understanding the circumstances surrounding the contract (as opposed to a mistaken belief held by both parties) will generally not allow the mistaken party to avoid a contractual commitment because its performance has become more difficult or even impossible.[2] Parties must also understand that their subsequent behavior and subsequent course of dealing has the potential to modify the terms of a written contract between them. In the United States, two separate legal principles can be used to contradict the terms of a contract or create a "quasi-contract" if a contract does not exist between the parties. In the first case, the theory of "promissory estoppel" entitles a party to recover from another party if the aggrieved party can show that the other party made a promise with the expectation that the promise would be relied upon by the performing party and that the performing party actually suffered damages and it was reasonable to rely upon the other party's promise. In the second case, the theory of "equitable estoppel" can be advanced as a means to compensate an aggrieved party to a contract that has been induced by a representation or act of its counterparty which statement or action was intended to induce (and, in fact, did induce) the other party to act to such party's detriment.

Parole (Oral) Evidence

Certain safeguards have developed over the centuries to protect parties who have had the sense to execute written agreements. The so-called "parole evidence rule" prevents one party to a written agreement from introducing into evidence any oral contractual term that conflicts with a written contract. However, it does not prevent a party from introducing oral evidence regarding the meaning or intent of a written term that is ambiguous in order to assist the court in assessing the parties' original intentions regarding the ambiguous written term. This policy of allowing a party to a written contract to introduce evidence of oral contractual terms can be particularly problematic in cases in which a contract fails to state explicitly that it encompasses the entire agreement between the parties and that there are no additional terms of the contract that are not contained in the written memorialization of the contract. In such cases, a party may enter evidence that there are oral additional terms that were not included in the written agreement so long as these terms do not conflict with any of the written terms of the contract. In these types of cases, in which one party alleges that there are additional oral terms of a contract, a court in a common law jurisdiction may look to the behavior of the parties to decide whether or not this is actually the case.

Additional Terms

If a party has notified the other party of its desire to include additional terms (such as those enumerated on the back of a purchase order) it is possible in a common law jurisdiction that the other party's failure to object expressly to these additional terms may result in the inclusion of these terms in the contract. These terms could possibly even supersede existing terms of a contract to the extent that they are inconsistent with the original terms. While courts in a common law jurisdiction are highly unlikely to construe a party's silence as its assent to new contractual terms, in order to avoid unintended results once the EPC contract has been executed, communications regarding the terms of the EPC contract should always be in writing. This writing itself should always note that it is not intended to modify any of the terms of the EPC contract unless it expressly states that it is intended to do so.

No contract can address all contingencies. No group of professionals can anticipate what circumstances may befall a project under development or construction. If something unforeseen arises (and it will) that is not squarely addressed in the EPC contract, a judge or an arbitrator in a common law jurisdiction can do nothing except look to the intent of the parties to hypothesize how the parties would have apportioned a risk had they been aware of it. Not surprisingly, the provision of an EPC contract that should be the cornerstone in its foundation is the clause most often missing entirely (or, if not, is usually hastily set in place in the recital clause without enough mortar to hold its intentions together to support the load of ambiguity it will be called upon to bear). It is the clause that explains the intent of the parties who are entering into the EPC contract. Too often the recital clause at the beginning of the EPC contract merely states that the sponsor desires to "retain the EPC contractor to engineer, procure and construct" the facility. Unfortunately, the clause usually does not go into any further detail of how the sponsor is expecting this feat to be accomplished. Should a costly problem arise that the EPC contractor does not want (and did not expect) to pay for, the EPC contractor's first line of defense will be that the work or problem at issue was not part of the scope of work called for by the EPC contract. A good example is an EPC contract that includes the demolition of a retired power plant (sometimes called a "dead soldier" in the industry) and its smoke stack to be accomplished by means of a wrecking ball so that sufficient space exists for the new power plant. If it turns out the old smoke stack was so well-built that only dynamite will work to knock down the smoke stack, who will pay for the additional cost could be left to be determined by the intent clause if the parties did not address this risk expressly in the EPC contract.

Without a well-constructed and comprehensive "intent" clause to take shelter under, a sponsor in a common law jurisdiction will be vulnerable to what justices who write court opinions like to refer to as the "great weight of authority," which mandates against imposing contractual obligations on a party (such as an EPC contractor) that a party has not expressly assumed under the contract that it has executed. The hallowed principle of "freedom of contract" permits parties to craft essentially any commercial arrangement that they desire, and a court or arbitrator in a common law jurisdiction will be reluctant to add a term to a contract (especially if it is inconsistent with another term) that the contract did not expressly contain on the presumption that the parties could have added the term but chose not to. In countries with civil codes, such as Chile, France or Germany, this is not the case. Contracts are subject to the relevant code involved. As a consequence of the approach of the common law legal system, a clause that summarizes the intent of the EPC contractor and the sponsor is essential.

In a common law jurisdiction, an "intent" clause should unambiguously recite the type of setting (for example, in the case of a power plant, an isolated grid vs. a large interconnected network, or, in the case of a mine, at high or low altitude) in which the facility will operate and the type of service that the facility will provide (for example, in the case of a power plant, peaking vs. baseloaded service). The intent clause should also state that the sponsor is expecting the EPC contractor to perform all work necessary to engineer, procure and construct a reliable facility that is designed according to the functional specification attached to the EPC contract and which can perform at all required performance levels, given prevailing environmental conditions, by the date that the EPC contractor has promised in the EPC contract. Under the EPC contracting approach, the EPC contractor is given a functional specification for the facility that the sponsor desires. There are many tasks and items of equipment that will be necessary for the EPC contractor to perform its assignment but not all of the tasks will be specifically identified in the functional specification. The intent clause must make clear that the EPC contractor must perform all tasks necessary to construct the facility in accordance with the requirements of the EPC contract, even though many tasks may not have been individually enumerated.

In many ways the sponsor's execution of an EPC contract is similar to a consumer purchasing a car from a car dealer. Once the consumer chooses the car model she desires and its engine size, she can expect a certain level of performance—speed potential, acceleration ability and miles per gallon. The consumer may order various "extras" as well. However, the consumer will not explain to the car dealer where to buy the parts to assemble the car or how to assemble them. The same is basically true for a facility such as a power plant constructed under the EPC approach. The sponsor will specify the engines, their output (capacity) and efficiency (heat rate) and also possibly add some extras (such as water injection [foggers] or dry low nitrous oxide emissions) but generally will not dictate how the plant must be assembled.[3] The "intent" clause of the EPC contract should make clear that the sponsor is not expecting to do any work concerning the power plant's design, procurement or construction unless the EPC contract expressly identifies the nature and extent of such work and the date by when such work must be completed.

Finally, as was discussed in Chapter 4, the intent clause should make clear that the EPC contractor should not rely upon any preliminary engineering analysis or design furnished to the EPC contractor by the sponsor. In most projects in which a sponsor has decided to employ the EPC approach, this reliance of the EPC contractor on preliminary design work should not be much of a concern. Typically, the sponsor will have completed only the most rudimentary of drawings and layouts so the EPC contractor can understand what type of facility the sponsor is expecting and, therefore, they will hardly be complete enough for the EPC contractor to use for anything other than a doorstop. However, in some cases, as in the instance of an extremely complicated project or an emergency project, the sponsor may undertake more than just preliminary design because it wants more control over what will be built as a result of its special project requirements or desire to elicit more accurate pricing and schedule terms (which will be possible if it provides potential EPC contractors with a more precise technical description). The project sponsor may begin to carry out but not complete the detailed design with the intention that the EPC contractor will take over and complete the design. The sponsor will try to gauge what portion of the design it has completed so that the EPC contractor has an idea how much work it has left to do. Levels such as "10 percent design" or "30 percent design" completed by the owner are fairly common depending upon the situation. This practice, often known as "bridging," can be quite risky for the project sponsor.[4] While courts generally uphold the all-encompassing, single point responsibility construct of the EPC approach, if that is what the parties intended,[5] courts may relieve the EPC contractor of some responsibility if the EPC approach was not in fact followed and the project sponsor assumed more responsibility than the sponsor was allocated under the EPC contract terms.[6]

As will be discussed in Chapter 8, the EPC contractor also should not be entitled to rely on any site-specific information furnished by the sponsor (such as the capacity of sewers and the gas pressures at which a pipeline is operating). The EPC contractor should be charged with the responsibility of investigating, or examining, all these matters on its own, so that it will not have the ability to claim that the sponsor was responsible for a shortfall in the facility's performance because the contractor relied to its detriment on information supplied by the sponsor.

Prudent Practices

Once the intent of the parties is made clear, to what standard of performance will the EPC contractor be held? What is the "baseline" or reference point for measuring the EPC contractor's performance? What measures can be used for purposes of comparison? There are industry codes and legal regulations that dictate what design and construction requirements the EPC contractor must employ for things such as pressure vessels, insulation, emissions and structural integrity. What about cases in which no code or regulation is directly applicable and the sponsor has not specified a particular standard or procedure? (After all, it is not possible for the sponsor to specify every last requirement and procedure.) To address this "gap" in the methods to carry out the work scope, the concept of "prudent practices" has developed. In the utility industry, for example, the use of terms such as "Prudent Industry Practices" or "Prudent Engineering Practices" or "Good Utility Practices" is an attempt to connote those practices that are adhered to by responsible engineering and construction firms in the utility industry. Thus, in the performance of its work, the EPC contractor will be judged by whether or not it has observed these "prudent practices." Since practices vary across regions and cultures, this standard is somewhat amorphous in practice. However, it is important that the EPC contractor and sponsor agree upon some definition of "prudent practices" and then expressly state that it is the intent of the parties that the EPC contractor will carry out the work in accordance with "prudent practices" so that an expert, mediator, arbitrator or judge will have a context and frame of reference in which to examine the EPC contractor's performance.

Sometimes terms such as "best practices" or "best available technology" are used. Most EPC contractors will resist the use of these terms because it may be difficult to determine when they have fulfilled their obligation since it means they must research the latest developments. Terms like these, however, are somewhat common when environmental remediation or issues are involved.

"Best" Efforts and "Commercially Reasonable" Efforts

In certain contexts, it is helpful to clarify what level of commitment a party must employ in carrying out its responsibilities under a contract. A categorical imperative, however, is frequently difficult to quantify. A good example of this might be the Supreme Court's opinion in the seminal case of *Brown v. Board of Education* in which Chief Justice Warren called for desegregation to be carried out "with all deliberate speed," whatever that means.[7] EPC contracts routinely bandy about such terms as "best efforts," "reasonable best efforts" and "due diligence" without expounding upon what is specifically intended by such terms (perhaps, because it is believed that these are legal terms of art, which is not true).

Contracts often set forth standards for a promisor's effort level required in order to attempt to fulfill its promise. Thus, terms such as "best efforts," "reasonable efforts" or "commercially reasonable efforts" are used. However, the distinctions between these terms are not clear.

Under New York state law, for example, is there any difference between the three standards of "best efforts," "reasonable efforts" and "commercially reasonable efforts" when such language is used in agreements? A provision governed by the "best efforts" standard is commonly construed to be the most onerous of the three standards.

However, this standard in New York law "is far from clear."[8] Courts have differed in how stringently to construe the "best efforts" standard but concur that "best efforts" is a term "which necessarily takes its meaning from the circumstances."[9] Therefore, a promisor is not expected to bankrupt itself to fulfill its covenant.[10] Whether a promisor has exerted its "best efforts," they seem to agree, is examined according to the promisor's own limitations.[11]

However, because the fundamental question is whether the promisor acted as a reasonable person would act in fulfilling a covenant, the line between "best efforts" and "reasonable efforts" is very vague.

In New York courts' application, failed reasonable attempts to fulfill an obligation are sufficient. For example, simply doing a reasonable search before making a business decision is sufficient to satisfy this standard.[12, 13] On the other hand, terms like "due diligence" tend to create confusion because they are used with different connotations in different areas of the law (such as the Securities Act in connection with defenses to liability for misrepresentations by underwriters, as was discussed in Chapter 4).

If the parties wish to specify, or even try to quantify, what level of effort is required, it is probably best to include examples of the behavior that is expected (for example, round-the-clock work shifts) or amount of expenditure required (for example, some percentage of the EPC contract price). If they are not able to agree, it may simply be useful to specify that the parties intend that the effort to carry out performance be commercially reasonable under the circumstances given the importance of the project to each party and each party's reliance on the other party to achieve the intended goal of construction of the facility.

"Good Faith" and "Good Intentions"

Many civil law jurisdictions impose a duty to negotiate in good faith. English law and law in most jurisdictions in the United States, however, do not impose such a duty. On the other hand, after a contract has been made, courts and statutes in many jurisdictions generally impose a duty of good faith and fair dealing upon parties in relation to their dealings pursuant to their contracts.[14] "Good faith" means "honesty in fact and the observance of reasonable commercial standards of fair dealing."[15] While this may sound reassuring, it may be quite difficult to show that a party has breached its obligations to act in good faith and has, in fact, acted in bad faith. Thus, this cause of action for breach of the duty of good faith is not very effective as a means of seeking recovery against a party that has not performed its end of a bargain. Courts may be loath to make a finding of "bad faith" because a party has not performed as required by the contract unless the injured party can demonstrate malice aforethought or recklessness.[16] Thus, while a party can be held accountable for its reprehensible behavior in the context of a contractual relationship and waivers of consequential liability in favor of an offending party probably cannot be used by a culprit to protect itself from economic retribution if it has committed deeds with malicious intent, a mere failure to perform by a party, without more, will generally not result in a finding of bad faith by a court.

Historically, punishment is a privilege that government reserves for itself. Hence, any claim of bad faith, with all the potential moral and representational ramifications that it engenders for a defendant, is one of the more difficult claims for a plaintiff to prevail upon unless the plaintiff can show some type of irreverently despicable behavior on the part of the defendant. It is a much repeated proposition in common law court opinions that "equity abhors a forfeiture" and, since a finding of "bad faith" is usually tantamount to a defendant's forfeiting all the contractual protections that it negotiated to protect itself, courts are generally reluctant to conclude that a defendant's actions have been undertaken in bad faith unless substantial and conclusive evidence is presented and corroborated at trial that the defendant has actually acted maliciously. As a corollary, courts will not condone a party's use of *in terrorem* (terror) practices or contractual provisions in order to compel another party's performance of its contractual obligations. As will be discussed in Chapter 17, penalties (as opposed to incentives) that punish and go beyond the extent of the injured party's economic harm will generally be voided by courts as against public policy because of their retributive nature (because retribution appears to be a realm that the state likes to reserve exclusively for its own dispensation).

In the heat of negotiations, when the hour is late and the conference room is stuffy and full of half-eaten sandwiches, the "intent" clause is rarely seen as a "hill to die for" when issues like price and liquidated damages are still open. This is an unfortunate perspective. Failure to hold out and put the flag on this hill may, one day, under improvident circumstances, cost the sponsor its financial return, its project or, worse yet, its company.

Notes

1 Most common law-based states in the United States have a "statute of frauds" that enumerates which types of contracts must be in writing to be enforced in court (such as contracts for the sale of real estate or contracts above a certain value or exceeding a certain duration). The statute of frauds is not intended to protect people who make oral contracts, but instead to protect people who did not make an oral contract and are accused of having done so.

2 See *Mistry Prabhudas Manji Eng. PVT. Ltd. v. Raytheon Engineers & Constructors, Inc.*, 213 F. Supp. 2d 20 (D.C. Mass. 2002), holding that both the EPC contractor and the owner must be mistaken as to expected performance for the doctrine of mutual mistake to apply.

3 One important exception to this is the case in which the sponsor is building (or already has) facilities similar to the one that it has retained the EPC contractor to build. In these cases, it is important to specify that the sponsor's goal is to minimize and optimize operations and maintenance costs. Consequently, the facility should be designed with systems similar to those in the sponsor's other facilities so that the sponsor's operators need only learn how to use or maintain one type of facility system, and their knowledge will be applicable to all the sponsor's other similar facilities. This will also make an economic difference in terms of

ordering, using, swapping and rebuilding spare parts. Without these instructions, the sponsor may be left with a facility that is incompatible with its other facilities because it has been constructed from a "hodgepodge" of leftover items from other jobs that the EPC contractor had undertaken for other clients.

4 See *Skidmore, Owings & Merrill v. Intrawest I Limited Partnership*, 1997 Wash. App. LEXIS 1505 (1997), holding an architect responsible for overruns caused by insufficient design specifications that a contractor used to support its bid for a construction project.

5 See *AgGrow Oils, L.L.C. v. National Union Fire Insurance Company of Pittsburgh*, 1A, 276 F. Supp. 2d 999 (S.D.N.D. 2003), in which it was held that a contractor had assumed full responsibility for all aspects of a seed processing facility.

6 See *M.A. Mortenson Co.*, ASBCA No. 53431, 03-1 BCA (CCH) ¶ 32,078, 2002 WL 31501914 (October 31, 2002), which held an owner responsible for design work it had supplied to a contractor. And see *United States v. Spearin*, 248 U.S. 132 (1918), in which a contractor building a dam was relieved of certain responsibilities when it built the dam according to the government's specifications.

7 *Brown v. Board of Education*, 349 U.S. 294, 301 (1955).

8 See *Bloor v. Falstaff Brewing Corp.*, 601 F.2d 609, 613 (2d Cir. 1979), with judges describing the differences between the standards as "quite murky"; and see *McDonald's Corp. v. Hinksman*, 1999 WL 441468, *12, 1999 U.S. Dist. LEXIS 9587, *36 (E.D.N.Y.).

9 *Perma Research & Development v. Singer Co.*, 308 F. Supp. 743, 748 (S.D.N.Y. 1970).

10 See *Bloor*, 601 F.2d at 615.

11 Ibid., at 613 ("It is obvious that any determination of the best efforts achievable by [promisor] must take into account [the promisor's] abilities and the opportunities which it created or faced").

12 *Brown v. Business Leadership Group*, 17 Misc.3d 1139, 856 N.Y.S.2d 22 (N.Y. Sup. 2007) (merely conducting a reasonable search to find a successor was sufficient).

13 See *Forward Industries v. Rolsn*, 506 N.Y.S.2d 453 (2d Dept. 1986), for a discussion of when a sponsor can expect best efforts from a contractor.

14 See *Hawken Northwest, Inc. and ADEC, J.U. v. State of Alaska, Department of Administration*, 76 P.3d 371 (Alaska 2003), noting that a "covenant of good faith and fair dealing is implied in every contract to give effect to the reasonable expectations of the parties." Also see § 1–203 of the Uniform Commercial Code, which provides that: "Every contract or duty within this Act imposes an obligation of good faith in its performance or enforcement."

15 U.C.C. §1–201(b)(20) (2012).

16 See *Union Carbide Corporation v. Siemens Westinghouse Power Corporation*, 2001 WL 1506005 (S.D.N.Y. Nov. 26, 2001), in which it was held that the mere allegations of "bad faith" supported only by evidence of a breach of contract could not survive the court's scrutiny as a matter of law and was dismissed.

Chapter 7

Subcontractors

No EPC contractor will be able to supply all the services and equipment necessary to build an infrastructure project. An EPC contractor will have to rely upon the skill and punctuality of many subcontractors and vendors. In the construction industry, the term "subcontractor" usually refers to someone who will perform work at the project site and the term "vendor" usually refers to someone who simply sells a piece of equipment to the EPC contractor or general contractor and otherwise has little involvement in the project other than perhaps giving technical advice regarding the installation or maintenance of its product. Subcontractors and vendors usually fall into two categories—those with whom the EPC contractor deals with on a regular basis (such as equipment manufacturers) and those whose expertise is necessary for the job at hand (such as local erectors and pipefitters) but with which the EPC contractor is unlikely to have extensive further dealings. See Figure 7.1.

The EPC contractor will enter into contracts with all these necessary parties, and one of the EPC contractor's goals will be to shift as much risk as possible to these subcontractors and vendors so that, if there is a problem (while the work is in progress or once it is complete), they will be responsible for any increased costs and delay. The sponsor has the EPC contractor "on the hook" for performance of the project and will seek recourse only from the EPC contractor and need not deeply concern itself with subcontractors and vendors. The EPC contractor, on the other hand, must protect itself by "passing through" its liability under contractual provisions of the EPC contract to its subcontractors and vendors so that the EPC contractor can try to protect its own profit.

As a consequence of the necessity for subcontracting, it is customary to permit the EPC contractor to subcontract work at its own risk. Most, if not all, EPC contracts contain provisions allowing the EPC contractor to employ subcontractors but provide that any such subcontracting will not absolve the EPC contractor from any liability whatsoever under the EPC contract. Obviously, the choice of inexperienced, insolvent or irresponsible subcontractors will ultimately be a problem for the sponsor because the EPC contractor ordinarily is not liable for delays and poor performance beyond certain agreed-upon monetary thresholds (as will be discussed in more detail in later chapters). In fact, insurance can generally be purchased by the EPC contractor to cover non-performance and bankruptcy of its subcontractors. As a result, sponsors and EPC contractors often agree upon a list of acceptable subcontractors and vendors in advance and attach this list to the EPC contract with a provision that the use of any subcontractor not on the list for work above a certain dollar amount or in a particular

Figure 7.1 Subcontractors.

discipline requires the sponsor's prior written consent, which, generally, the sponsor agrees not to unreasonably withhold, condition or delay. The EPC contract should note that the failure of the sponsor to approve a subcontractor should not be treated as a *force majeure* event that excuses the EPC contractor from performing.

The EPC contractor will usually agree to deliver to the sponsor copies of all major subcontracts so that the sponsor, if it wishes, can check them for consistency with the EPC contract's provisions and scope. Often, the sponsor will agree that the EPC contractor may "redact" all pricing information from subcontracts that are delivered to the sponsor (especially because the subcontract may contain confidentiality provisions regarding disclosure of the price in the case of a complicated piece of equipment). While this redaction is probably not necessary because the sponsor has already agreed to pay a fixed price for the work by signing the EPC contract, the EPC contractor may not want the sponsor to see the real "mark-up" that the EPC contractor has placed on the subcontractor's price, because the sponsor may believe that the mark-up's magnitude is inappropriate. Unless the sponsor believes that it is being taken advantage of by the EPC contractor, there is little reason that the sponsor will need to see the price of each subcontract. In fact, the sponsor will probably be in a similar position to that of the EPC contractor if the EPC contractor requests to see a copy of the sponsor's offtake or use agreement for the project with the project's offtaker or user. The sponsor is unlikely to be eager to share the offtake or use pricing with the EPC contractor because the EPC contractor may very well be able to estimate the profitability of the sponsor's project and keep this sum in mind if the EPC contractor must later request reimbursement for any costs incurred if there are changes in the work during the project. This is one reason why, as will be seen in Chapter 12, unit pricing for changes in work should be set and included in the EPC contract. In summary, neither the sponsor nor the EPC contractor generally needs to see the economic relationship with the other party's counterparties, and arriving at an understanding on this point early in the negotiation process is probably in the interest of both parties.

Sponsors are also well-advised to negotiate the right to receive prompt and written notice from the EPC contractor if any significant changes or claims are made under any of the EPC contractor's major subcontracts. This type of notice can often serve as an early warning signal to the sponsor with respect to issues that the EPC contractor may be ignoring or underestimating and that can be handled before they escalate into problems. Minimum thresholds for the notification requirement can be set (in terms of cost or delay) so that the EPC contractor does not have to notify the sponsor of inconsequential problems.

Sponsors should understand that an EPC contractor often has significant leverage over its subcontractors because these subcontractors may be involved in other projects that the EPC contractor is building. On the other hand, the EPC contractor may be reluctant to share or contractually prohibited from sharing copies of its subcontracts with the sponsor. For the sponsor, there is a delicate balance that must be struck between vigilance and interference, especially since the sponsor does not have "privity" of contract with subcontractors (meaning that the subcontractors and sponsor are not parties to the same agreement and do not have any contractual relationship).

Zealous observation of the performance of subcontractors (especially those on site) will stand the sponsor in good stead because it will allow the sponsor to identify non-conforming work (which is best corrected while it is still in progress). Contrarily, supervision and direction of the work by a sponsor can lead to successful claims by the EPC contractor that the sponsor, not the EPC contractor, is responsible for any shortfalls that emerge in the project's performance or any circumstances that delay the project's commercial operation as a result of the sponsor's assumption of control of the project (see Chapter 10). Usually, the best approach for the sponsor is to negotiate rights to observe and comment on work while it is in progress but to be cognizant that it should not instruct the EPC contractor as to how to proceed with the work—merely convey to the EPC contractor what the sponsor has identified as issues of concern.

Not all issues are capable of being resolved in a friendly discussion or exchange of letters because substantial time and cost can underlie items in contention. For this reason, as will be discussed in Chapter 12, it is necessary that the EPC contract contain safeguards for the sponsor that give the sponsor the absolute right to direct changes in the work unilaterally if the sponsor, in its sole discretion, deems such changes necessary. Who will bear the cost of these changes will usually be the subject of (a sometimes acrimonious) debate in a conference room or a courtroom at a later date. Generally, the most serious errors in a project occur during its design phase, not its construction. This is true because some mistakes in design cannot be fixed during construction. The only solution for a design miscalculation is demolition and re-design. Otherwise, the facility may never function correctly. For example, in a power plant, if its fuel (coal) handling system is incorrectly designed and not enough railcars can be unloaded in the time required, the power plant may never reliably operate at the electrical output that the sponsor expected. Only re-design could remedy this type of problem. Therefore, it is much better to identify such shortcomings before construction commences.

Problems in the construction phase can be serious, of course. Suppose the sponsor observes that the concrete slab upon which the turbines of the power plant will rest has been improperly poured and "cold joints" have formed in the foundation because the concrete did not dry properly. The owner may believe that the slab will not last for the intended life of the power plant. In opposition, the EPC contractor may claim that the slab is structurally sound as poured (bearing in mind that the EPC contractor usually gives a one- or two-year warranty on its work, although many states have statutes that impose a 10-year or more liability period on civil structures, and these are often referred to as "decennial liability statutes"). In all of these situations it is necessary that the sponsor contractually have the right to direct the outcome it desires (which may include tearing up and re-pouring the entire slab). As will be discussed in later chapters, at whose expense this work will be performed is another issue that may have to be determined later by an independent expert, an arbitrator or a judge.

Terms Contained in Subcontracts

EPC contracts usually contain provisions mandating what types of terms must be included or excluded from the EPC contractor's subcontracts. It is common to attempt to provide that subcontracts be assignable to the sponsor without the consent of the subcontractor in the event that the EPC contractor defaults in its performance under the EPC contract. (Assignment of contracts is discussed in Chapter 25 and default is discussed in Chapter 13.) The rationale for assignment of subcontracts to the sponsor is that the sponsor can assume all the subcontracts and thus complete the project with minimum delay and expense if the EPC contractor does not perform its obligations. While this assignment provision appears reassuring, in practice it is unlikely to be very effective. If the EPC contractor is in difficulty, financial or otherwise, it is likely that the subcontractors have not been paid and have stopped work or even "walked off the job" and terminated or simply abandoned their subcontracts with the EPC contractor as a result of non-payment. Realistically, if the sponsor desires to employ the subcontractors or a new EPC contractor (which rarely happens because retaining a new EPC contractor would likely entail significant expense and delay), the sponsor will probably need to negotiate new subcontracts with these subcontractors. In addition, as will be discussed in Chapter 21, sponsors often require the EPC contractor to prohibit its subcontractors from filing any claims or liens against the owner or the facility. Again, in reality, if a subcontractor is at the point at which it is prepared to file a claim against the EPC contractor, it will usually disregard any provisions of its subcontract and also file claims against the owner of the facility for work performed. However, unless there is a statute entitling the subcontractor to recover from the sponsor, it will generally not be able to win such recovery.

Third-Party Beneficiaries

Attempting to sort out all such issues in all the EPC contractor's various subcontracts is probably not a productive use of the sponsor's resources. Instead, the sponsor should focus its attention on the EPC contractor's subcontracts for key items and should probably require that all major subcontracts (that is, those above a certain dollar value or those concerning critical equipment) include the protections mentioned above and even require subcontractors that are party to these major subcontracts to deliver acknowledgments to the sponsor that the sponsor is an intended "third-party" beneficiary of these major subcontracts so that the sponsor can rely upon the provisions of these major subcontracts if a problem develops.[1]

It is also important that major subcontractors be creditworthy or obtain adequate performance bonds if they are not. Creditworthiness also may often concern the sponsor's financiers, so the sponsor must be ready to field such inquiries from its financiers. In many cases, as a result of short supply or long lead times, the sponsor may have already entered into its own contracts directly with suppliers of major equipment such as turbines and transformers. When the EPC contract is signed, the sponsor may assign these contracts to the EPC contractor at the time notice to proceed is given, and, if so, they will become subcontracts. In this event, it is crucial for the sponsor to obtain a release of all liability from these suppliers at the time these supply contracts are assigned to the EPC contractor so that the sponsor is not caught in any disputes regarding this contract or the subcontractor's work.

It is also prudent to provide that, if the sponsor terminates the EPC contract for its convenience (to be discussed in Chapter 15), the sponsor will have the right to "re-assume" these assigned supply contracts if the sponsor wants to keep the equipment that these suppliers were to furnish. However, bankruptcy of the EPC contractor might make it difficult for the sponsor to enforce its rights under a "reassignment" provision unless the suppliers themselves have entered into direct contractual agreements with the sponsor agreeing to this reassignment because, unfortunately, equipment supplied to the EPC contractor by these suppliers is likely to be deemed part of the "estate" of the bankrupt EPC contractor once these supply contracts become subcontracts.

Note

1 A third-party beneficiary of a contract is a party that is not a signatory to the contract but is a party that the signatories to the contract intend to permit to rely upon that contract as if it were actually a party to the contract. Unless a party has specifically been named as a third-party beneficiary of a contract by the parties to a contract, it is fairly difficult, but possible, for a party to prove that it was intended to be a third-party beneficiary of the contract in question. See *Port Chester Electrical Construction Corp. v. Atlas,* 40 N.Y.2d 652, 655 (1979). Under the laws of England, there is a common law doctrine of privity of contract that provides that only parties to a contract will have directly enforceable rights thereunder—see *Dunlop Pneumatic Tyre Co Ltd v. Selfridge & Co Ltd* [1915] UKHL 1. There are a number of statutory exceptions to this common law doctrine: the most relevant in the construction context is the Contract (Rights of Third Parties) Act 1999 (Third Parties Act). Pursuant to the Third Parties Act, parties to a contract may grant an "identified" and "identifiable" third party the right to enforce certain terms. Commonly, these third-party rights are included in a stand-alone schedule or a schedule detailing the relevant contractual provisions that a third party may enforce. The Third Parties Act applies but parties may expressly exclude its application to their contract.

Chapter 8

Site or Route Survey

No project can be properly designed or constructed without a careful analysis of the conditions at its site or on its route. Meteorological and topographical conditions are often well documented and a matter of record. Disappointingly, the same is not true for subsurface conditions. While the subsurface state of the site or route can sometimes be surmised from local empirical data because other excavation has been carried out near the site or on the route, only full-scale earthwork will reveal what, if any, obstacles to construction are present below the surface. Both urban and rural areas can pose problems. In cities, buried streams, utility lines and old foundations are just some of the obstacles that can be unexpectedly uncovered. In less populated areas, underground boulders, caverns, unstable rock and natural wells can all complicate a job that was expected to be routine. Because unexpected conditions lead to delay and expense, a site or route survey should be undertaken well in advance of commencing any work in an attempt to anticipate what conditions will be encountered. The question of when, by whom and how this survey is done depends upon the commercial circumstances involved. In fact, in some projects such as highways, railroad and subways, a detailed survey of the route is often not feasible at all if the terrain is rough or the route is lengthy.

Who bears the risk of unexpected subsurface conditions often depends upon who has chosen the site or route. If a government or offtaker has dictated the sponsor's use of a particular site or route, the government or utility will usually agree to be responsible for any costs or delay associated with subsurface conditions. If, however, the sponsor has selected the site or route, often it (and not the government or offtaker) will bear the risk of subsurface conditions. In general, legally, as regards EPC contractors and sponsors, EPC contractors and not sponsors bear the risk of unknown subsurface conditions at the site unless the EPC contract expressly provides otherwise by means of what is usually referred to as a "differing conditions" clause, which serves to shift the cost of differing site conditions back to the sponsor.[1]

More often than not, however, the EPC contractor shifts the burden of unanticipated subsurface site conditions to the sponsor in the EPC contract. The EPC contractor will often argue that, if this risk is not assumed by the sponsor, the EPC contractor will have to add a contingency to the EPC contract price to account for the possibility of conditions that differ materially from those expected. This is a potentially costly contingency to add to the EPC contract price for the sponsor, given that it may never materialize. While the EPC contractor's argument that it will add a steep contingency to the EPC contract price in order to cover unexpected subsurface conditions may have some truth in the context of "sole source" negotiations (as were discussed in Chapter 4), this "added contingency" argument may lose some of its validity in the context of the RFP approach in which bidders are competing against one another and may look to be competitive by reducing pricing contingencies in their price in order to try to win the bid.

In most cases, well before an EPC contractor is selected, the sponsor conducts a boring survey of the site or route to try to detect any serious subsurface impedances. Once an EPC contractor has been selected, some preliminary design layout is done and the general location and footprint of heavy components is determined, a more thorough subsurface boring investigation can be undertaken. Timing usually dictates how this boring investigation is handled. If the sponsor is under intense time pressure (such as an emergency power supply project), usually the sponsor (through a consultant) will undertake a site survey before the EPC contract is executed. The sponsor will often take responsibility for any conditions not disclosed by the site survey.

If, however, the sponsor is not in such a rush, the sponsor is well-advised to require the EPC contractor to conduct the site or route survey either before or after the EPC contract has been signed. The sponsor and the EPC contractor should agree in writing upon what assumptions the EPC contractor has used in formulating its EPC price (for example, number and depth of foundation piles planned) before the site or route survey is conducted. Then, the EPC contractor should be given time to conduct a site or route investigation according to a protocol and scope reviewed in advance by the sponsor. The EPC contractor should then be given a further period after the site or route survey is complete to assess results of the site or route survey and submit for the sponsor's consideration any changes to its construction assumptions necessitated by results of the survey. If the sponsor agrees, the EPC contract price can be changed on the basis of added costs revealed by the survey. If parties cannot agree upon the costs of the changes (which rarely happens), the EPC contractor will be paid for work associated with the survey, notice to proceed will never be given, and the parties will go their separate ways. EPC contractors should remember to include a provision in the EPC contract providing for payment for the survey whether or not notice to proceed is ever given by the sponsor.

If a change to the EPC contract terms or price has been agreed by the parties (or if no change has been requested by the EPC contractor within the time allotted), the EPC contract should provide that the EPC contractor will not be entitled to any cost reimbursement as a consequence of additional costs it incurs as a result of any subsurface conditions encountered as work progresses. EPC contractors commonly resist such preclusion of relief. Often, the EPC contractor is able to include provisions in the EPC contract that will grant the EPC contractor additional schedule or cost relief if the EPC contractor subsequently encounters a subsurface condition that would not have been discovered by a reasonable survey conducted by a responsible surveyor familiar with the conditions in the vicinity of the site or route and who had acted in accordance with prudent industry practices in conducting its survey. Whether or not the EPC contractor is able to negotiate inclusion of this more lenient provision in the EPC contract, it is, however, very typical for the EPC contractor and sponsor to agree that the EPC contractor's discovery of certain items such as antiquities, foundations or other man-made objects (which site surveys are generally not designed to detect) will always entitle the EPC contractor to an extension of time and/or increase in the EPC contract price if the EPC contractor will incur increased costs or delays as a result of its discovery.

It is also helpful to provide a descriptive and illustrative list of what sort of conditions will not entitle the EPC contractor to relief once the site survey is complete and the EPC contractor's timeframe in which to submit requests for relief has lapsed. This list often includes soft soil, caves, boulders and underground rock formations. Another good approach is to agree with the EPC contractor on the type of foundation that the EPC contractor intends to use and to allow the EPC contractor relief if the foundation assumptions are not in fact appropriate for the conditions. For example, the EPC contractor may estimate that a 4-foot thick foundation with 20 pilings of 18 feet in depth may be needed. Of course, site conditions may eventually turn out to be different than expected so perhaps 35 pilings of 60 feet each are actually required. If this is the case, a pre-agreed schedule of price per additional piling and per additional foot of pile can avoid an unnecessary dispute claim. Similar approaches can be developed for tunneling work based upon rock class expected to be encountered and water seepage expected to flow from the tunnel cut. The sponsor should also require the EPC contractor to make a representation in the EPC contract that the EPC contractor has carefully examined and is familiar with all aspects of the site including ingress, egress and its elevation.

Hazardous Materials

Rarely will the EPC contractor agree to assume the risk of discovery of hazardous materials on the site or along the route. Although local laws can differ, testing for contamination is usually not required to be performed at a construction site or on a route unless the existence of hazardous materials can be reasonably suspected by the parties (for example, in the case of a brownfield site that hosted an environmentally dangerous application). While EPC contractors will almost never take responsibility for the time or cost involved in disposing of, or entombing, hazardous materials, if the sponsor suspects that certain materials may be discovered (such as petroleum-contaminated soil), it should try to negotiate contract provisions that require the EPC contractor to conduct the work involved in remediating such hazardous materials on a time-and-materials basis. The sponsor should try to include unit and labor rates for remediation items in the EPC contract. Making the EPC contractor responsible for remediation (but not its cost) can serve to avoid lengthy negotiations between the sponsor and the EPC contractor if hazardous materials are, in fact, discovered. It is best to make the EPC contractor responsible for handling and coordinating any hazardous material clean-up because the EPC contractor is usually in a better position than the sponsor to arrange for clean-up and integrate clean-up into the project schedule with minimum delay. Employing a separate contractor to remediate hazardous material under the direction and control of the sponsor is not advisable (as will be seen in Chapter 10). Delay and interference claims from the EPC contractor regarding that contractor's work are much more likely to be successfully lodged against the sponsor if the EPC contractor suffers additional costs or delays while the EPC contractor awaits the remediation of the hazardous materials.

Legal Control over the Site

Whether the sponsor purchases, leases or enters into a usufruct for the project site often is inconsequential in terms of the sponsor's ability to control the site and obtain financing so long as the jurisdiction permits the sponsor to mortgage its interest in the site and appurtenant rights of way to its lenders as collateral for their loans.

In terms of risk allocation, leases and usufructs often provide for termination of certain breaches such as non-payment or failure to pay taxes and these types of risks will not exist if the sponsor purchases the site outright. On the other hand, should the site be or become contaminated, the sponsor may be more successful in shifting the risk of remediation and harm to the site owner through indemnification, contribution or merely abandonment or vacation of the site. In many jurisdictions such as the United States, being a lessee as opposed to an owner will usually do little if anything to insulate the sponsor from governmental claims concerning pollution.

Note

1 See *P&Z Pacific, Inc. v. Panorama Apartments, Inc.*, 372 F.2d 759 (9th Cir. 1967), holding that a contractor was responsible for the costs of moving boulders that were not detected in subsoil boring reports. See also *Pinkerton and Laws Co., Inc. v. Roadway Express, Inc.*, 650 F. Supp. 1138 (N.D. Ga. 1986), ruling that a subcontractor was responsible for costs involved in encountering moist soil.

Chapter 9

The Work

Once the EPC contract has outlined the paradigm under which the parties are operating—that it is their intent that the EPC contractor work in accordance with prudent industry practices to construct a reliable facility that meets the sponsor's requirements—it is necessary to elaborate on the work more specifically. While it is the general idea that the EPC contractor will furnish everything necessary to complete the facility whether or not an item is expressly specified in the EPC contract, the EPC contractor will need guidance in terms of the sponsor's expectations as to how the work will be performed and how it expects its facility to function.

The Functional Specification

This brings the parties back to the "functional specification" of the facility referred to in Chapter 4, which the sponsor initially used to evaluate the economics of the project and solicit bids from EPC contractors. This document will now become the basis for what is usually called the "scope of work," "work scope," "scope," "specification," "functional specification," "technical specification," "performance specification," "tech spec," "design criteria" or "design specification." It will describe the facility that the sponsor desires. This functional specification will serve as an outline for the EPC contractor to follow in order to design and construct the facility.

As mentioned in Chapter 4, the functional specification is not meant to be a compendium of every task and specification involved in the project but rather a description of the functions and characteristics that the sponsor intends the project to possess once it is built. As a matter of nomenclature, it is probably best not to use the term "scope of work" or "work scope" or "scope" for this document. Use of the term "scope" seems to imply there is work outside of the EPC contractor's scope of responsibilities in connection with completing the project—a dangerous implication that has the potential to counteract the basic premise of EPC contracting (which posits that the EPC contractor will be responsible for all work necessary to complete the project unless the EPC contract expressly states otherwise).[1]

In any case, the more specificity the sponsor includes in the functional specification as to design, equipment and work methods, the more likely the sponsor will not be able to hold the EPC contractor accountable if the facility does not perform as the sponsor intended because the EPC contractor may well be able to point to the sponsor's plans and drawings as the source (or at least a contributing factor) for the facility's inadequate performance. As a result, it is common for sponsors to include statements in the EPC contract to the effect that the EPC contractor has performed its own evaluation and investigation of the items of reference supplied by the sponsor and that the EPC contractor has not relied upon information supplied by the sponsor in its bid for, or prosecution of, the work. However, most frequently these broad statements are of little use when an EPC contractor can point to a detailed design drawing that was developed by the sponsor.[2] In representational terms, a functional specification should be thought of as a schematic representation of the sponsor's facility.

Consequently, equally as critical as agreeing upon the commercial terms of the EPC contract is the process of transforming the initial draft of the functional specification into an integrated functional specification.

A good functional specification will provide a lucid conceptual description of the facility the sponsor desires, the work practices that are acceptable to the sponsor, the parameters to be used in the EPC contractor's supply of equipment for the facility and the procedures for testing the facility's equipment.

Too often, since many different specialties and disciplines are involved in preparing the functional specification (ranging from civil and mechanical engineering to electric and chemical engineering), different professionals prepare different sections of the functional specification. A table of contents is simply tabulated for these sections and the functional specification is considered complete. This can lead to disaster for the sponsor because inconsistencies and even omissions may emerge once the EPC contractor begins its work. It is crucial that the sponsor charge one of its own engineers with the task of reviewing all the sections of the functional specification for completeness, consistency with the other parts of the EPC contract and clarity. In general, while EPC contractors can be held responsible for their failure to point out and inquire about patent ambiguities in a document that the sponsor has supplied, problems that are not blatant are usually the sponsor's responsibility.[3]

Once the functional specification has been refined and agreed upon by the EPC contractor and the sponsor, it will be included as an exhibit to the executed EPC contract. Typically, the body of the EPC contract itself, in addition to the functional specification, will contain provisions governing the EPC contractor's performance of the work as well, so it is the job of the sponsor's lawyers and technical advisors to ensure that there are no conflicts between terms in the body of the EPC contract and the EPC contract's functional specification exhibit. Because problems do not always become apparent until work begins, it is common to provide for an "order of precedence" of the different sections of the EPC contract (that is, the body and the exhibits and appendices) in case confusion emerges. This order of precedence will determine which specification, instruction or procedure will prevail if a discrepancy among different sections of the EPC contract exists.

Contractual Provisions

While the functional specification exhibit of the EPC contract will contain the technical details of the work, the provisions of the EPC contract itself will generally contain a narrative description of the nature of work that the EPC contractor is expected to undertake. A well-drafted EPC contract will note that the EPC contractor must perform all work necessary to complete the sponsor's facility including the work described in the functional specification as well as any other work that is implied by either provisions of the EPC contract, the functional specification or prudent industry practices. The EPC contract should explain the types of activities the EPC contractor is expected to perform in execution of the work. Generally, either the provisions of the EPC contract or the functional specification should expand upon each of the areas outlined below.

The EPC Contractor's Obligations during the Design Phase

Professional Engineering

The EPC contractor should be required to employ only skilled and licensed professionals in connection with the design of the facility to ensure that the facility as designed and as contemplated to be operated will comply with all relevant industry and building codes.[4] The facility must be designed to take into account expected seismic activity in its area. In addition to earthquakes and tremors, seismic activity can lead to "liquefaction" in which the strength and stiffness of saturated soils is reduced over time and unexpected settlement can occur. Inadequate foundations and anchoring can lead to permanently debilitating/operational restrictions such as bullet trains being unable to run safely at their design speed and coking ovens being unable to be charged with their full coke load capacity.

The facility should be designed not to interfere with neighboring commercial or industrial activities during its construction or operation. The EPC contractor must be responsible for familiarizing itself with all laws relating to the facility and be held accountable for any remedial actions or violations (including fines) that are a consequence of its failure to conform the work to any of these requirements. Noise can be a good example of an area in which the EPC contractor complies with all regulatory requirements but local farmers or residents nonetheless raise claims that lead to bad will or lawsuits if attenuation measures (such as sound barriers) are not implemented by a project sponsor. Sponsors will generally implement measures to try to placate the local community even if they are not legally obligated to do so because no sponsor wants to create a bad relationship with the local community. All these measures will be carried out at the sponsor's expense unless it has been careful to impose on the EPC contractor a higher standard than simply the often nebulous or vague maximum decibel requirement of applicable regulations (which often do not specify how to take into account background noise or specify where and how many noise receptors must be employed to test noise and vibration levels). Typically, however, while the EPC contractor is responsible for conforming the work to any unexpected changes in laws or industrial codes that occur after the signing of the EPC contract, it is the sponsor that will usually agree to pay for any increased costs associated with these changes in law or industrial codes after the EPC contract has been executed on the premise that such cost could not have been taken into account by the EPC contractor at the time the EPC contract was entered into.

Quality Control

The EPC contractor should be required to implement a quality control program acceptable to the sponsor and designed to detect errors and omissions in the work promptly so that they can be corrected as early as possible. The International Standards Organization (ISO) rates entities by certification level. The sponsor can use these ratings as an objective evaluation of the EPC contractor's qualifications in different disciplines such as welding, or working at high heights or in confined spaces. The quality assurance program should continue throughout design, construction and testing of the project. It is important for the sponsor to monitor this in "real time." For instance, are the welders up to the job? The sponsor should know this before the job is 70 percent welded.

The EPC Contractor's Obligations during the Construction Phase

Health and Safety

Of paramount importance is the safety of life and property. The EPC contractor should be required to put in place a safety program for all workers (including those of subcontractors working at the site or elsewhere) designed to eliminate undue risk of accidents. Unchocked truck tires, day workers sleeping at the worksite at night and failure to investigate driving violations are all good examples of prescriptions for unnecessary work interruption. A safety coordinator should be appointed and site-specific emergency procedures should be developed and coordinated with local health, fire and law enforcement authorities so that all parties will be prepared to act in a coordinated manner if an accident occurs.

Supervision

The EPC contractor should be required to maintain adequate supervision of the work and maintain good order at the construction site or any other site where work is to be performed. The EPC contractor should be required to and the owner should have the absolute right (without the EPC contractor being entitled to claim *force majeure* or owner delay) to eject any person from any worksite who endangers any person or property.

Project and Construction Managers

The EPC contractor should submit the credentials of the candidates whom it proposes to act as the project manager and construction manager (which the sponsor may agree can be the same person if the project is not too complex), at the signing of the EPC contract a list of the individuals who are acceptable to the sponsor should be attached to the EPC contract, and the EPC contractor should be required to designate the individuals from the list who will serve in each position once notice to proceed is given. (Depending upon the size and complexity of the project, the owner may even want to add more positions to the list, such as engineering manager or chemical process manager.) The EPC contractor should not be allowed to change the individuals in these positions unless they leave the EPC contractor's employ. The owner should not agree that it will not unreasonably withhold its consent to a substitution of these individuals because that can lead to disputes. The EPC contractor should find individuals who are suited, both mentally and psychologically, to the task at hand. For instance, a remote site on a lengthy project requires a certain personality and the EPC contractor must choose its staff carefully. Changing individuals after a project begins can only lead to unnecessary disruption of a project. In order to avoid this, the sponsor may want to put some "teeth" into this provision. Otherwise, if the EPC contractor falls behind on another project or its project manager becomes homesick, instead of the EPC contractor offering a raise or other incentive to deal with the situation, the EPC contractor may just try to find a new project manager. The owner may want to impose liquidated damages on an EPC contractor that changes the project or construction manager. If liquidated damages may not be legally enforceable in the relevant jurisdiction, then the intent clause of the EPC contract should note that the owner is contracting for the skill of the particular individuals named as the project and construction manager and a discount on the EPC contract price will be granted if those individuals do not fulfill their roles throughout the term of the project. Provisions like these are key in industries (like mining) in which experienced managers tend to be in short supply because the industry has slumped or is nascent and therefore there is a paucity of seasoned professionals (who can often easily change jobs because generally they are always receiving attractive employment offers).

Union Labor

It is desirable to reach an express understanding in the EPC contract with the EPC contractor as to whether or not union labor will be required by the sponsor at the construction site. Often, a union labor requirement (which should be applicable to all subcontractors as well) will minimize the likelihood of delays occasioned by strikes or other labor incidents at the construction site or in the project's vicinity. Unless the EPC contract has expressly shifted responsibility for strikes to the sponsor, under the law of most U.S. states, labor unrest, labor shortages and similar occurrences will generally not excuse the EPC contractor from its obligations under the EPC contract. However, it is quite typical, especially when strikes are national or regional as opposed to site specific, that the sponsor (usually under the *force majeure* provision of the EPC contract as will be discussed in Chapter 12) will grant at least a schedule extension and sometimes cost relief for these incidents, because they are unlikely to have been provoked or predicted by the EPC contractor.[5]

Security and Surveillance

The EPC contractor should be required to maintain security of the construction site by enclosing or patrolling its perimeter and preventing unauthorized persons from entering the worksite. The EPC contractor should assume responsibility for theft and vandalism and expressly waive any claims for delay, costs or *force majeure* against the owner arising out of any such incidents.[6]

Minimization of Disruptions in Surrounding Areas

The EPC contractor should be required to submit a work plan that demonstrates that adjacent properties and their operations will suffer minimum (if any) adverse effects as a result of construction activities, particularly in the case of additional water or road traffic congestion as a result of the EPC contractor's work. The EPC contractor should also be required to observe all local ordinances related to work and traffic restrictions (for example, not working or accepting deliveries during the night or on weekends).

Coordination with Other Contractors

The EPC contractor should be required to coordinate its activities with the activities of any other contractors or parties involved in the project (such as the utility crews who will interconnect to the plant each of the gas, power, telephone and water lines) in order to avoid unnecessary delays and costs, such as delays that can result from the need to handle and move arriving equipment more than once. Such coordination can avoid a situation such as a vendor delivering heavy equipment to a site before the concrete foundations for the equipment have been set. Otherwise, equipment must be stored and moved again. "Double handling" is costly and creates unnecessary risks.[7] So-called "coordination agreements" between all contractors who will work at the site and the sponsor are a common practice in some industries such as the windfarm power sector (where a turbine supplier usually will deliver and install its wind turbine generators but another contractor will supply the rest of the civil work [such as access roads and foundations] and electrical work [such as power lines and the substation and transformer] and the two contractors must coordinate) or in an integrated project such as a steel mill complex, which may have different EPC contractors for different parts of its facility such as the port, the coking oven, the power plant, the mill and the railyard.

Site or Route

The EPC contractor should be obligated to maintain a relatively clean site or route so that hazards do not develop and neighboring property owners do not complain to local authorities.

Shipping and Delivery Schedules

The EPC contractor should be required to prepare a schedule of expected shipping and delivery dates for major shipments so that, as will be seen in Chapter 23, the sponsor and its insurers can satisfy themselves that the transportation process (including the modes of transportation and the carriers themselves) is adequate and reasonable given the project's circumstances at the time (for example, whether the project is ahead of or behind schedule or whether local, domestic or international political concerns mitigate against a particular shipping route, port of call, or carrier's flag). The sponsor should also have the right to inspect equipment and materials shipments to see that equipment and materials have been properly packed. An insurer may deny coverage of a valuable shipment if it has not been able to inspect a cargo before the vessel leaves port, and it may be impossible to inspect the cargo at a subsequent port of call if the ship's captain will not allow its entry to the ship or if the cargo container to be inspected is buried below many other, lighter containers because the captain will not want to waste time and pay the port demurrage fees for the ship's standing in port beyond its allotted docking window.

Sourcing of Equipment

Occasionally, it is necessary (as a result of government incentives or taxes or restrictions such as embargoes or boycotts) for the sponsor to mandate or forbid that certain equipment or services required to complete the facility be procured (or not procured) from certain jurisdictions. This may be done to preserve the sponsor's fiscal benefits or eligibility for financing. It may also be done to avoid contractual penalties imposed by the offtaker or host government if the sponsor has not sourced a minimum proportion of its equipment and services domestically. Requirements like this are common in industries such as solar power and rail in which the government may be trying to incubate an industry such as solar modules or signaling and communications. Domestic requirements sometimes require that even domestic insurance and financing be obtained by the sponsor. In these cases and also for taxation and accounting reasons, it is usually propitious for the sponsor to request that the EPC contractor and its subcontractors deliver certificates of origin in connection with all equipment and services rendered to the sponsor. For instance, the sponsor might have obtained financing for some of the project's major equipment from the export credit agency of a foreign government that is offering such financing to promote its indigenous equipment industry.

The EPC Contractor's Training Obligations during Commissioning

During the commissioning period, the EPC contractor should be required to provide a complete training program for the sponsor's operating personnel. The sponsor should note specifically what type of classroom and field training it expects the EPC contractor to provide and whether it will be videotaped or recorded for future use. The sponsor must also decide and communicate what level of detail and description (and in what language) it expects in the operating manuals that the EPC contractor will prepare.

Establishing the Framework

An EPC contract that creates an appropriately detailed framework for the EPC contractor's responsibilities yields many benefits. First, it will help the EPC contractor understand what the sponsor expects and, thereby, allow the EPC contractor to make an informed assessment of its cost for performance of the EPC contract. Second, as will be seen in Chapter 11, it will help both the EPC contractor and the sponsor measure progress of the work. Then, if the sponsor is not satisfied with the EPC contractor's progress or performance, an objective measure will have been established for the sponsor, the EPC contractor, a technical expert, a mediator, an arbitrator or a judge to assess asserted deficiencies.

Notes

1 See *Wild-Fire, Inc. v. Laughlin, et al.*, 2001 Ohio App. LEXIS 976 (March 9, 2001), in which it was held that a contractor's claim for damages for out-of-scope work was denied because the contractor failed to demonstrate that the work performed was beyond the scope of work contemplated by the parties. But see *J.T. Majors, Inc. v. Lippert Bros., Inc.*, 263 F.2d 650 (10th Cir. 1958), holding that "cleaning" of bricks was not necessarily implied by a contract calling for the "installation" of bricks.

2 See *Edsall Construction Co., Inc.*, ASBCA No. 51787, 2001–2 BCA ¶ 31,425, in which the Armed Services Board of Contract Appeals held that a contractor could recover additional costs incurred when the government had furnished an inappropriate design specification for the design that the government had requested. The government attempted to contest its liability by claiming that it had actually provided a performance specification and not a design specification because the government had included a disclaimer of liability in the documents that required contractors to verify the assumptions that the government had made in preparing the performance specifications that had been given previously to the contractor.

3 See *Crowley, Inc. v. United States*, 923 F.2d 871 (D.C. Cir. 1991), holding that the notation "R-value 12.5" was unclear as to whether or not each insulating layer was required to possess such R-value or all the layers cumulatively were required to possess such R-value.

4 See *Hess v. Tube Zone Realty Co.*, 94 N.J.L. 4 (1924), holding that a contractor that agreed to erect buildings in compliance with municipal law could not avoid its responsibility to the owner when the local municipality rejected the contractor's plans. "The contract being express to provide a thing not in itself unlawful, the defendant was under an obligation to perform it, notwithstanding the fact that the cost of performance would have been very much greater than it anticipated at the time it made the contract."

5 See *Curtis Electric Co., Inc. v. Hampshire House, Inc.*, 142 N.J. Super. 537 (Law Div. 1976), relieving a contractor of liability for a strike based on its specific inclusion in a *force majeure* provision but noting, by way of *obiter dictum* (non-binding observation of the court), that "generally delays occasioned by strikes do not excuse the non-performance of contracts."

6 See *Sandwick, Inc. v. Statewide Sec. Systems*, 192 N.J. Super. 272 (App. Div. 1983), holding that a security service could not be excused from liability when its security guard was bribed to leave the site being guarded so that a robbery could be perpetrated. Under several different legal theories such as *respondeat superior*, an employer can be held responsible for the failure of its employees to perform their duties and employers will generally only be relieved of liability when employees act outside of the scope of their duties (for example, such as committing fraud and criminal acts, etc.).

7 Oddly enough, while the sponsor usually cannot be held responsible to the EPC contractor for the work of other contractors at the site if the sponsor has included a "no damage for delay" clause, as will be discussed in Chapter 14, unless the sponsor has expressly assumed an obligation to coordinate contractors, the contractors can usually sue one another for the costs of their delays. See *Broadway Maintenance Corp. v. Rutgers*, 90 N.J. 253 (1982), holding that the sponsor was not liable to various contractors for delay but the contractors had recourse against one another pursuant to a contractual provision that provided that any contractor would be liable for delays and costs that it caused any other contractor.

Chapter 10

The Sponsor's Obligations under the EPC Contract

Aside from its obligation to pay the EPC contractor, as will be discussed in Chapter 21, the sponsor will be responsible for relatively little else under the EPC contract except, perhaps, providing some of the various permits, services and equipment that the EPC contractor will need to construct the facility.

Customarily, the sponsor will obtain emissions and discharge permits such as permits for air emissions, effluent discharge and storm water runoff. The EPC contractor will usually obtain all permits relating to construction activities and occupancy. The sponsor should attach a specific list of permits that it will obtain and require the EPC contractor to support its efforts to obtain them. The EPC contract should specify clearly that any and all other permits concerning the facility's design or construction (not operation) will be applied for and obtained by the EPC contractor, whether or not they have been specifically enumerated in the EPC contract. Although the EPC contractor may resist this type of all-encompassing responsibility for permits, the sponsor should point out to the EPC contractor that the sponsor is not in the construction business and is retaining the EPC contractor to carry out all work involved in construction of the facility, including design and construction-related approvals and permits. Realistically, since the EPC contractor will usually be engaging local subcontractors to carry out most of the on-site work and these local contractors will usually be familiar with legal requirements placed upon them by local authorities, the EPC contractor is in a better position than the sponsor to ascertain these requirements.

The sponsor may also be responsible for providing some of the physical interconnection facilities and related utility services, as and when such services are required. The sponsor should set expected dates for the availability of these services and include them on the baseline project schedule so the sponsor does not become obligated to make these services available earlier than expected and thereby risk a claim of owner delay. For example, it is always helpful to delineate the arrangements for any electricity that will be required prior to facility completion. Generally, the EPC contractor will be required to supply and pay for any electricity needed to support construction activities (either through a connection to the local utility or on-site generators). In the case of a power plant, for example, electricity necessary for commissioning and testing must come from the power grid because on-site diesel generators for construction equipment power cannot produce the electricity needed. This "backfeed" power comes from the electricity grid. For example, transformers and electrical switchyards are "soaked" with electricity from the power grid (generally for 24 to 48 hours) before energy from the power plant is sent through them. In general, the sponsor will arrange and pay for this "backfeed" power. In fact, most power plants and processing plants, unless they have so-called "black start" capability (for example, batteries or on-site diesel generators), cannot be started unless they are receiving "backfeed" power from the grid.

Finally, the EPC contractor will usually ask the sponsor to agree that the sponsor and any of the sponsor's own contractors and suppliers (such as the utility interconnection providers) will not delay the EPC contractor's work. The sponsor should be alert to ensure that this type of provision is related only to its own actions (or omissions) and those of its own agents and contractors and never agree to language that is much more expansive in its scope such as "delays not attributable to the EPC contractor," which is a much lower legal hurdle for the EPC contractor to jump over than "not caused by the owner" language in court or in an arbitration. Otherwise, the sponsor may inadvertently assume responsibility for the delay of a party that the sponsor expected the EPC contractor to bear, such as for a freight forwarder. As will be discussed in greater detail in Chapter 14, as a legal matter, the sponsor will generally not be responsible for any delays that it causes in the EPC contractor's work unless the sponsor has "actively interfered" with the EPC contractor's work or the sponsor has acted unreasonably in preventing the EPC contractor from prosecuting its work.

Assumption of Responsibility

While the EPC contract will provide that the sponsor will only be responsible for the duties assigned to the sponsor in the EPC contract, the sponsor must be careful that its subsequent behavior is not inconsistent with this provision. If the sponsor subsequently begins to handle a matter that is not within the sponsor's specified scope of duties under the EPC contract, whether or not this matter was specifically allocated to the EPC contractor under the EPC contract, may become irrelevant if the EPC contractor relies upon the sponsor's handling of the matter. In such cases, while one might say that the sponsor has been "deputized" by the EPC contractor, courts will generally hold that the sponsor has simply volunteered to perform the activity in question. Consequently, the sponsor may be held accountable to the EPC contractor for any of the EPC contractor's costs and delays that result from the sponsor's failure to perform its newly assumed duties effectively. These unfortunate situations typically arise when an EPC contractor encounters difficulty in dealing with a third party, such as a permitting agency issuing a construction-related permit or a property owner from which the EPC contractor must obtain a right of way or easement (such as for access to the site or lay-down or parking area for equipment and vehicles). If the sponsor chooses to become involved in such situations (which it may have good reason to do), the sponsor should make certain that the EPC contractor acknowledges in writing that the EPC contractor will waive any claims that the EPC contractor might have against the sponsor as a consequence of the sponsor's assumption of these additional responsibilities.[1]

Owner Furnished Equipment

Occasionally, the sponsor may purchase (or may have previously purchased) some of the major equipment for the facility. In the case of power plants or facilities requiring their own power such as an LNG gasification terminal, this is particularly true if gas turbines are to be employed (which the sponsor may have bought in quantity for several of its upcoming projects in order to secure a discount or a delivery date or in the "aftermarket" from a developer whose project has been aborted or postponed).

In instances where the sponsor is supplying some of the project's equipment, there are several scenarios. If the equipment has yet to be built, the sponsor can simply assign its equipment purchase contract directly to the EPC contractor and (with permission of the vendor) relieve itself of all liability under the equipment purchase contract. In this case, if a problem arises with the vendor's performance or its equipment, it will be the EPC contractor's problem, and the sponsor will be able to call upon the EPC contractor to solve the problem. It will then be up to the EPC contractor to pursue any recourse that it may have against the equipment vendor. On the other hand, if the equipment has already been built and, perhaps, is sitting in a warehouse, it may be easier for the sponsor to assign certain rights (such as warranty and installation assistance) under the equipment purchase contract to the EPC contractor but retain ownership of the equipment. Otherwise, the sponsor would have to transfer ownership of the equipment of the EPC contractor (through a bill of sale in the United States) and then the EPC contractor would have to "resell" the equipment to the sponsor under the EPC contract. This type of transaction could trigger sales taxes and a mark-up on the equipment price from the EPC contractor for the administrative costs that the EPC contractor will incur in connection with the transfer. Also, theoretically, this is a risky approach for the sponsor in the sense that the equipment could become part of the EPC contractor's estate if the EPC contractor were to go bankrupt between the time that the EPC contractor purchases the equipment from the sponsor and the time that the EPC contractor resells the equipment to the sponsor.[2] This approach gets more complicated if the sponsor has made only partial payment to the vendor and the vendor has not yet fully assembled the equipment. In the case of fully fabricated equipment that is being stored in a warehouse, issues such as who will be responsible for costs if it turns out to have been improperly maintained during its storage or damaged during storage, or if its manuals and specifications have been lost since it was placed into storage, and loading, shipping and unloading of the equipment will arise because a delay in these activities can result in a delay in the EPC contractor's other activities and ultimately an obligation on the part of the EPC contractor to pay delay liquidated damages (as will be seen in Chapter 17) if the EPC contractor does not complete the facility by the required deadline.

Refurbished Equipment

The situation can become still more complicated if already fabricated equipment must be retrofitted or restored to meet project requirements. The contractor who will refurbish the equipment, usually the original equipment manufacturer (OEM), is not likely to agree to charge a fixed price for refurbishment because the OEM will usually not be able to determine precisely what will be involved in the refurbishment until the OEM disassembles the equipment so that its internal condition can be inspected. The inability to rely upon a fixed price for equipment refurbishment can result in lenders requiring the sponsor to raise the amount allocated for contingencies in the project's budget. Thus, the equity amount that the lenders will require the owner to commit to invest in the project may increase accordingly.

Lenders will generally require the sponsor to increase the contingency amount set aside in the project budget if a contingency actually arises before they begin lending. This is based on the logic that the contingency line item in the project budget has been set aside to pay for unknown contingencies, and once an item has arisen, a separate allocation should be made for it in the budget because it is really no longer an unknown "contingency." If this practice were not followed, the amount set aside in the budget as a reserve for unknown contingencies would actually decrease. Finally, equipment may have been in storage so long that the OEM's warranties may have expired or been invalidated if the equipment has been modified. The OEM may charge a hefty fee to renew its warranty in such cases.

While the purchase of "pre-owned" or even used equipment might lower the capital cost enough to justify its use, the sponsor must be wary of the unexpected issues this approach can cause during implementation of the project. To the extent that the sponsor keeps (or the EPC contractor refuses to accept) any risks that would normally be borne by the EPC contractor if it were procuring new equipment as part of its scope of work, the sponsor's lenders may choose to revise the terms upon which they are willing to lend to the sponsor. For instance, as mentioned above, they may require the sponsor to increase its equity commitment in the project to make certain that any cost overruns associated with refurbishment of the equipment will be paid for by the sponsor. While refurbishing equipment may initially appear to be a cost-effective option, this option may actually require a proportionately larger capital commitment from the sponsor who is using non-recourse project financing to build its project relative to the use of "new and clean" equipment and, therefore, increase the sponsor's financial exposure and probably decrease the sponsor's return on its investment because it will have the same revenue but will have made a larger investment.

Technology Licenses

Some facilities (like polypropylene plants or refineries) rely on a particular proprietary technology that is central to their successful operation. If this is the case, generally the owner will sign a long-term license agreement directly with the technology owner so that the owner can be assured that the owner has all the rights that it will need to exploit its facility and need not rely on the EPC contractor to obtain these rights (given their importance to the project). In addition, these licenses will usually provide the owner with assistance and update services throughout the lifecycle of the facility. Such arrangements generally require that the EPC contractor obtain and pay for its own license with the technology owner to fabricate and install the proprietary technology. (See Chapter 22 for a further discussion of licenses.)

Notes

1 See *Commonwealth of Pa., State Highway & Bridge Auth. v. Gen. Asphalt Parking*, 405 A.2d 1138 (Pa. Commw. Ct. 1974), ruling that an owner was liable to a contractor for delays involved when the owner assumed responsibility for negotiating the relocation of a water main.

2 If the sponsor has purchased the equipment before setting up a special purpose subsidiary to carry out the project, the sponsor should be aware that the most tax-efficient method of transferring the equipment to its newly created subsidiary may be to contribute this equipment as a capital contribution to the newly formed subsidiary or even merge the sponsor's company that purchased the equipment into the newly created subsidiary, which is a "tax-free" reorganization under the Internal Revenue Code in the United States and may have similar tax treatment in other jurisdictions.

Chapter 11

Monitoring the Progress of the Work

Once "notice to proceed" has been given to the EPC contractor, the sponsor would be ill-advised to go back to its office to await a phone call from the EPC contractor announcing the project's completion. Constant vigilance by the sponsor is required throughout the design and construction of the project.

Developing the Baseline Project Schedule

Without a good work schedule, confusion and delay are inevitable. Before the EPC contract is signed, the EPC contractor should produce a detailed schedule outlining the sequence and duration of each work activity. This "baseline" project schedule should be attached to the EPC contract as an exhibit when the EPC contract is signed. The sponsor should scrutinize this project schedule to confirm that the contractor has anticipated and incorporated all the important events that will occur during design and construction. The EPC contractor might not be aware of some of these events because the sponsor has failed to communicate them to the EPC contractor (such as, in the case of a power plant, LNG terminal or refinery, the fact that the nomination period notice has been given to the fuel supplier so it can arrange or reserve gas or oil field production and transportation).

The project schedule should identify all critical path activities (that is, any activities whose delay in commencement or completion will delay the project's in-service date). If reasonably possible under the circumstances, it is desirable that the project schedule be calibrated based upon days elapsed since notice to proceed has been given instead of actual calendar dates (which, of course, must be premised upon the actual calendar date that notice to proceed is expected to be given). This will be helpful in the event that notice to proceed is not given on the calendar date anticipated by the parties, which is often the case, and, thus, there will be no doubt about the impact of a delay in the sponsor's giving notice to proceed on the calendar date that the parties had in mind. It is also helpful for the project schedule to specify by when third parties, such as permitting agencies, shipbuilders and interconnection utility providers, must complete their tasks in order for the EPC contractor to maintain the project schedule. If the EPC contract provides that the EPC contractor is required to seek the sponsor's specific review of any part of the work, the sponsor and the EPC contractor should always agree upon how much time the baseline project schedule should allot for such review (and appropriate time for at least several iterations of comments) and incorporate these timeframes into the baseline project schedule.

While the actual project schedule may change constantly as circumstances change, the baseline project schedule never changes (other than the one exception mentioned below). The baseline schedule stays "frozen" forever so that the project's progress against the parties' original expectations can be measured. Often, however, the EPC contractor will be permitted to submit a revision to the baseline project schedule during the first few weeks after the EPC contract has been executed in order to take into account such things as the lead times for supply of items that the EPC contractor proceeds to order once the EPC contract has been signed (these items generally are not ordered by the EPC contractor before the EPC contract is signed because there is always the possibility that the sponsor and the EPC contractor will not sign the EPC contract and, thus, the EPC contractor might forfeit the down payments that these supply contracts often require) or site or route conditions that have been discovered during the site or route survey (as was discussed in Chapter 8). If, after evaluating the EPC contractor's proposed revisions to the baseline project schedule, the sponsor agrees with the EPC contractor's requests, a new baseline project schedule will replace the original baseline project schedule.

Schedule Updates and Progress Reports

The EPC contract should require that the EPC contractor revise the project schedule on a periodic basis (most typically monthly). The revised project schedule should be submitted to the sponsor in a format that can easily be manipulated by the sponsor, and the sponsor should specify the precise hard copy and software format that it desires the EPC contractor to use to make sure such software is compatible with the EPC contractor's systems. If the sponsor's lenders or offtaker will require the sponsor to deliver schedule updates to them, the sponsor should remember to confirm that the format they will require will be consistent with the format that the EPC contractor will be delivering to the sponsor. This project schedule can be compared to the baseline project schedule to track progress and identify emerging problems. For instance, the baseline project schedule may have allowed three weeks for soil settlement before construction, but soil may still be settling six weeks after its compaction.[1] Thus, the project schedule will often change. The EPC contractor should be required to explain why changes from the baseline project schedule have occurred, their impact and how the EPC contractor plans to mitigate them, and also what, if any, activities have migrated onto the critical path. The last item is especially important because some activities that originally were not on the critical path can become critical path activities if their commencement or execution is delayed. For example, if a cargo is lost at sea and, therefore, equipment must be re-ordered (or, worse yet, fabricated), an equipment delivery that was not initially on the critical path could easily become a critical path item.

Rights of Inspection

Engineering and Design Phase

As noted in Chapter 7, most critical decisions regarding a project's construction, cost and performance will occur during its design. A project will be constructed based upon decisions made during the design phase. For this reason, the sponsor is well-advised to review all important drawings and documents produced during this period so that it can make suggestions or raise any concerns it might have. From a legal point of view, although it may appear to be a picayune distinction, it is desirable for the sponsor to be entitled to review and comment upon the EPC contractor's design work rather than actually to approve or reject the EPC contractor's work product. If the sponsor has approved a plan and then the facility does not perform as the EPC contractor has warranted, the EPC contractor's classic defense to the sponsor's assertion of violation to the warranties will be that the sponsor approved the (deficient) design and, therefore, the EPC contractor should not be held accountable for any diminution in performance of the facility that resulted from implementing the (deficient) design. It is wise for a sponsor to try to protect itself from such allegations by the EPC contractor by including language in the EPC contract that provides that the sponsor's approval of, or concurrence in, any plan or drawing submitted by the EPC contractor shall not be deemed acceptance of the work nor be deemed to relieve the EPC contractor of any responsibility under the EPC contract to achieve all the performance levels for the facility guaranteed by the EPC contractor in the EPC contract. This statement should unequivocally confirm for any tribunal that is construing the EPC contract that it was not the intent of the parties to the EPC contract to shift any responsibility for inadequate design of the facility to the sponsor. However, in court or an arbitration proceeding, this provision, unfortunately for the sponsor, may not actually yield the intended protection for the sponsor if it has expressly "approved" a mistake or error or omission in the facility's design because the sponsor's behavior will have been inconsistent with the contractual provision, which the tribunal may, therefore, disregard. In practice, sponsors, if they want to stamp drawings, should stamp them "no further comment" or "no comment" but not stamp them "approved."

Thus, the requirement that the sponsor approve design documents should be replaced with the right of the sponsor to comment on design documents. Ideally, the EPC contractor should be required to resolve, or at least address, any of the sponsor's design commentary to the satisfaction of the sponsor, and the EPC contractor should not be permitted to proceed with any work based upon a design that is in question until the sponsor is content with the resolution of any issues concerning such design. Of course, any delay occurring as a result of this iterative comment process can expose the sponsor to a claim of delay from the EPC contractor. This is a risk that can be mitigated (but not eliminated entirely) by providing for discrete periods in the baseline project schedule during which the sponsor must provide comments on design documents to the EPC contractor and further deadlines by which the EPC contractor must respond to the sponsor regarding these comments. Generally, it is to the EPC contractor's advantage (in order to be able to maintain the project schedule) to make sure that the EPC contract clearly states that the sponsor's failure to provide reasonably detailed written comments during the time allotted for comments will be deemed conclusive evidence that the sponsor has no objections to the design work in question.

On a practical level, a list of the design work that will be required to be submitted by the EPC contractor to the sponsor for its review will often be compiled and attached to the EPC contract as an exhibit. While this is in the EPC contractor's best interest so that there is no confusion, it is not always in the sponsor's best interest because the sponsor may overlook an item that is not on the list or simply not be aware of an item, and then the sponsor may forfeit the right to review this item later when the sponsor realizes that it has been omitted from the list. The sponsor should try to retain the right to review any design work that the sponsor requests whether or not the design item appears on the list of documents listed in the EPC contract to be submitted by the EPC contractor to the sponsor for review. EPC contractors will generally oppose this type of "open-ended" right to review any design document and maintain that they are not able to share "proprietary" items, such as shop drawings, with the sponsor (or even any party) because such items may, in fact, be trade secrets. In response, the sponsor can suggest that design documents, such as shop drawings, relating to components that the EPC contractor purchases from vendors need not be disclosed but the facility system design documentation prepared by the EPC contractor and its subcontractors for the sponsor's facility should be made available for the sponsor's review.

A sponsor may have differing levels of concern regarding design review depending upon the type of facility being constructed. For instance, in the case of a power plant, a simple cycle gas or solar plant generally comes in modular components from vendors that are interconnected at the construction site. Consequently, the overall design work to be done by the EPC contractor may not be so complicated or extensive. In fact, most combustion turbines, particularly so-called "aero-derivative" turbines whose design has evolved from jet engines, are designed so that they can be transported by railcar to facilitate rapid delivery and installation.[2] On the other hand, a coal-fired plant requires special design of a boiler, which arrives in pieces and must be fabricated at the construction site, and usually special design of a steam turbine as well based upon the boiler size and steam pressure. Thus, solid fuel burning power plants are not "off the shelf" in any sense and carry much more risk that improper design will result in failure to meet the target performance. Not surprisingly, in this case, the sponsor may have a heightened level of concern. Usually, a fair compromise with the EPC contractor is to allow the owner to maintain the right to review any plans that, in the owner's opinion, could be expected to have a material impact on the performance of the facility.

Manufacturing and Fabrication Phase

Most equipment and components will be manufactured and assembled in factories far from the construction site. Typically, components such as transformers are usually tested at the place of their fabrication before they are shipped to the site. Given the expense involved in monitoring the manufacture of all of the various components involved in a project, the sponsor usually will rely upon the EPC contract's warranty provisions to deal with failure of any particular component. Of course, failure of a component during the warranty period can mean an outage of the facility from service—a potentially costly interruption in the sponsor's operations and profits, which will not be covered by the EPC contractor's warranty (but may be covered by business interruption insurance once a deductible period has elapsed—see Chapter 23). Thus, major equipment components are often inspected by the sponsor while they are being assembled.

Witness Points vs. Hold Points

Many fabrication facilities employ production-line assembly methods, and fabricators are loath to disrupt their production sequencing. The sponsor and the EPC contractor are usually given the right to observe certain stages of assembly on the fabricator's shop floor while fabrication is in progress. The fabricator's subcontract with the EPC contractor often will identify and provide the EPC contractor with advance notice of specified events in the production cycle so that they may be witnessed to ensure proper quality and quantity without interfering with the fabricator's overall production process (this is known as a "witness point"). On the other hand, assembly of a component may be so critical or special that the fabricator is actually required to hold up production until an inspection or test is accomplished (this is known as a "hold point"). The fabricator is not allowed to continue its work until the EPC contractor (or sponsor) is satisfied that the work is adequate.

Field Construction

Not surprisingly, most disputes seem to arise over work that is done at the construction site. Miscommunication, lack of coordination, imprudent construction practices and unexpected meteorological or geological conditions are just some of the circumstances that give rise to defective work and accidents, all of which cost time and money.

Sponsors are usually good at overseeing construction progress because construction takes place in an easily observable setting. Unfortunately, sponsors are usually not diligent about documenting and transmitting their complaints or criticisms to the EPC contractor in writing during the construction stage. Their failure to do so usually leads to the perpetuation, rather than elimination, of construction disputes.

Worksite productivity can often be a problem. Generally, there is little a sponsor can do about this problem except badger the EPC contractor and rely on provisions of the EPC contract, which should require that the EPC contractor submit a remedial plan if it falls behind schedule and accelerate its performance by working more shifts if the EPC contractor is behind schedule. Often, though, the EPC contractor will not accelerate the work if it believes that it can meet the scheduled in-service date or that it will be less costly to pay some amount of delay liquidated damages to the sponsor rather than pay overtime wages. As will be discussed in Chapter 13, short of commencing a lawsuit seeking "adequate assurance" from the EPC contractor that the EPC contractor will perform, there is, practically, little the sponsor can do unless it wants to try to terminate the EPC contract and find a replacement contractor. One way to try to keep an EPC contractor on schedule is to include so-called "interim" milestones in the EPC contract (for example, foundations poured by a certain date, major components to arrive at the site by certain dates, etc.) which the EPC contractor must achieve or pay liquidated damages for missing any of these interim milestones. However, EPC contractors generally will not agree to include interim milestones in the EPC contract because they want to be free to carry out the work in any sequence they desire, especially if problems arise that they need to work around, and if they do agree they usually require a complete refund of interim delay liquidated damages if they actually achieve the target in-service date. A good compromise in this situation is to assess the liquidated damages based upon interim milestones but refund or decline to collect them if the EPC contractor achieves the promised in-service date. The best way to guard against schedule delay is selecting a responsible EPC contractor with a good track record of on-time completion of its projects because preventing delays is the most vulnerable support in the EPC contract's foundation.

In terms of rectifying improper work, the sponsor has more options. The sponsor should negotiate the right to inspect (and even dismantle) work in progress if it desires to determine whether or not work is defective. This disassembly should be at the EPC contractor's cost if the work is found to be defective or if the EPC contractor failed to notify the sponsor in time for the sponsor to exercise any observation rights that the sponsor negotiated in the EPC contract.

Occasionally, disputes will arise if the sponsor objects to an item or method that is not specifically mentioned in the functional specification but which, in the sponsor's opinion, violates the intent of the EPC contract or prudent industry practices. Perhaps the EPC contractor is not, in the sponsor's opinion, using sufficient signage measures for the protection of local traffic or not using "ripwrap" to secure shorelines or employing sufficient dust suppression techniques. Of course, if the EPC contractor believes that it is acting according to prudent industry practices (usually defined in the EPC contract as those practices employed by reputable firms engaged in the relevant industry in the geographic region in which the project is located [see Chapter 6]), it will often simply continue the work and threaten the sponsor with a delay claim if the sponsor continues to interfere.[3] In these cases, it is usually wise for the sponsor to retain a reputable engineering expert to study the situation and deliver its findings in writing promptly to the sponsor. The sponsor should deliver these findings to the EPC contractor. But, as was discussed in Chapter 4, the sponsor should consult its attorneys as to the form in which this information should be provided so that the attorney/client privilege will not be compromised. This procedure will establish a record that may facilitate an informal resolution of the matter or will serve to prepare the matter for a formal dispute proceeding.

Project Accounting and Audit

In connection with monitoring work progress and documenting a record for the future, the EPC contract should require the EPC contractor to maintain, both during the project and for an agreed-upon period after project completion, adequate books and records concerning the project that the sponsor should have the right to audit (at the EPC contractor's expense if any material discrepancies are found). EPC contractors often resist this audit right on the basis that their work is performed on a lump-sum basis at the EPC contractor's sole risk if the work is more costly than anticipated. This argument has some merit and therefore, often, audit rights are limited to audits in connection with change orders not being performed on a lump-sum basis, allowing the sponsor to verify the costs for which it is reimbursing the EPC contractor.

The "Owner's Engineer"

Frequently, the sponsor will retain an engineering firm to assist with supervision of the EPC contractor and administration of the EPC contract. This so-called "owner's engineer" or "project manager" or "construction manager" will give the sponsor expert advice on progress of the work and the EPC contractor's performance. Although EPC contractors are usually leery of the owner's engineer, a competent owner's engineer will often help facilitate a successful project from both the owner's and the EPC contractor's points of view.

The "Independent Engineer" or "Lender's Engineer"

If lenders are involved in financing the sponsor's project, they will appoint an expert engineering firm (whose fees will typically be paid by the sponsor) to advise them on matters relating to the budgeting, engineering, procurement and construction of the project. While usually referred to as the "independent engineer," this expert is not at all independent but works to advise the lenders and protect their interests, which may or may not be aligned with those of the sponsor depending upon the matters involved. Occasionally, differences of opinion arise between the independent engineer and the sponsor (or the owner's engineer) because the independent engineer will try to maximize the project's operational reliability and minimize risk for its client (the lenders), while the sponsor will generally be attempting to balance the elimination of risk against the incurrence of additional capital and operating costs. Consequently, arguments can arise between these parties over redundancy (for example, one pump vs. one pump plus a back-up pump or two smaller transformers [at a higher price] vs. one large transformer [at a lower price]) or which spare parts are useful (but costly) to purchase and keep on-site (so-called "strategic" spares, which run the risk of becoming obsolete or simply never needed).

Notes

1 See *American Family Mut. Ins. Co. v. Pleasant Co.*, 268 Wis. 2d 16 (2004), which addressed this very issue when a building foundation cracked as a result of soil settlement and the building had to be demolished.
2 An aero-derivative turbine can often be "swapped out" in a matter of hours and replaced with a spare engine while it undergoes repairs. However, since these aero-derivative engines are usually very efficient and produce relatively little exhaust heat (often called "waste heat") compared to larger turbines, they are of somewhat limited use in combined cycle operation.
3 See *WPC, Inc.*, ASBCA No. 53,964, 04-1 BCA ¶ 32,476 (2003), in which an owner and a contractor disputed whether "shotcrete" was cast-in-place concrete and therefore could be used for a project. Also see *Strouth v. Pools by Murphy and Sons, Inc.*, 79 Conn. App. 55, 829 A.2d 102 (2003), affirming the right of an owner to terminate a contract with a contractor because the contractor was not building a swimming pool according to the contract's specifications.

Chapter 12

Unforeseen Circumstances, Adjustments and Change Orders

Historically, once a party had signed a contract, generally even the impossibility of performance could not excuse the contracting party from being liable for damages if it failed to perform its obligations under the contract. U.S. courts generally refuse to excuse performance based on grounds of unexpected expense or unanticipated delay. Parties to contracts are free to negotiate contractual provisions regarding circumstances that will excuse performance. If the parties have chosen not to include an excuse from performance in a particular circumstance, a court will usually require a party to perform its contract. After all, the party in distress was free not to enter into the contractual obligation in the first place. Thus, U.S. courts have tended to construe contractual provisions relating to excuse from performance narrowly unless it is unequivocally clear that the intent of the parties was to excuse performance as a result of the impediment involved.

Force Majeure

During the nineteenth century certain common law doctrines (such as mutual mistake and *force majeure*) developed that served to excuse counterparties from their obligations in the case that their performance was impossible, either because they had both made a mistake in their contracting or as a result of an act of nature (often referred to as an "act of God"). These doctrines were construed narrowly and certain occurrences could excuse performance only if their consequence was, on an objective basis, such that no person (not just the non-performing contract party) could perform the obligation under the contract.[1]

For example, if a contractor agrees to build a five-story building and a storm destroys the building once the contractor finishes the third story, under common law, no *force majeure* has occurred. However, if a contractor is hired to install an elevator in a five-story building and the five-story building is destroyed in a storm, that is *force majeure* under common law.

Because unexpected circumstances arise, contracts for the provision of goods and services customarily contain *force majeure* provisions that excuse a party from its obligations to the extent that a "force" beyond its control has prevented its performance. However, *force majeure* never excuses payment obligations.

A well-drafted "excuse" or *force majeure* provision will describe the type of circumstances that will excuse a party from performance but provide that the party will be excused from its performance only to the extent that the circumstance could not have been prevented by the party claiming the relief if such party were using reasonable commercial efforts to prevent or overcome the circumstance. Thus, it is in the sponsor's interest that the description of the circumstances for excusal of the EPC contractor's performance be very specific and not general. Ideally, circumstances should be listed specifically or, in the alternative, classified by type. Classifications such as "events beyond a party's control" or "acts of God" should be avoided because they are unnecessarily and ambiguously broad and should be replaced with specified occurrences such as "earthquakes, tornadoes, and other similar natural environmental disasters not involving human intervention." In general, events such as strikes and governmental action (such as embargoes and sanctions) will not be construed by courts as events that excuse performance unless they are specifically listed as events that excuse performance in the EPC contract. Actions of social unrest such as strikes and protests are often included as *force majeure* events so long as they are regional or national and not directed solely or specifically at the affected party in connection with its practices in carrying out the project work.

A well-drafted *force majeure* provision will also contain an illustrative list of what types of events and circumstances should not be considered *force majeure*. Usually excluded from *force majeure* are instances of "economic hardship" such as changes in market supplies or prices, labor shortages, and deterioration in the financial condition of a party. In fact, it is interesting to observe how the litany of *force majeure* events enumerated in EPC contracts often varies from country to country and region to region around the world. The *force majeure* clause is often an exposé of the geopolitical reality of the project's location, so sponsors may want to study the "contents" of the *force majeure* provisions carefully in order to help assess the risks involved in a particular project's location. *Force majeure* provisions of some EPC contracts may contain an expansive description of insurgency and terrorism, provisions in other EPC contracts may focus on union and labor relations, while yet other provisions in EPC contracts may enumerate ubiquitous local geographical or meteorological conditions such as difficult site access in certain seasons or soft soil.

Thus, a carefully drafted *force majeure* provision will contain a two-pronged test in determining whether or not an event is eligible to be granted *force majeure* status. First, the event must fall within an enumerated category of events. Second, the affected party must not have been able to avoid the consequences of the event once it occurred. In practice, it is typical to relax this standard of requiring the affected party to be unable to overcome the impact of the event by simply requiring the affected party to have used "reasonable commercial efforts" to overcome the event's occurrence or consequences. This dispensation is usually given because many things can be prevented if enough money is spent. (See Chapter 6 for a discussion of "reasonable efforts.")

Occasionally, a third prong is added to the test that requires that the event not have been reasonably foreseeable by the affected party at the time the EPC contract was entered into by the parties. This requirement is based upon the presumption that the affected party could have taken adequate precautions to avoid the event's occurrence if the event was foreseeable. Often the inclusion of this "foreseeability" limitation will engender significant discussion at the negotiating table. Without a doubt, many things (such as lightning strikes and high winds) are foreseeable because they happen. Often, a good compromise is to set a benchmark for "foreseeability." For example, weather conditions beyond those typically and historically encountered at the site (as evidenced by weather service records) might be grounds for *force majeure* relief, but seasonal flooding or rain during the rainy season probably should not be grounds for *force majeure* relief. The EPC contractor should be obliged to plan around predictable occurrences such as rainy seasons and integrate their expected consequences into the baseline project schedule (as was discussed in Chapter 11). A *force majeure* provision should actually state how many days of weather-related delay each month have been incorporated into the baseline project schedule and not permit any *force majeure* relief for the EPC contractor for inclement weather until that number of days of bad weather has elapsed in that month.

The *force majeure* provision should also state how many days of "float" are in the baseline project schedule and note whether or not *force majeure* relief should be granted while there is still "float" in the current project schedule. For instance, assume that a 460-day baseline project schedule provides for 10 days of "float" or "slippage." Only 450 work days are actually necessary to complete the project but 10 days have been added to the baseline project schedule as buffer. What if, by day 200, the EPC contractor has "consumed" nine days of the float due to a shipping accident and, on day 201, heavy rain prevents work at the construction site for five days? Should the EPC contractor be entitled to claim *force majeure*? If so, should it be entitled to four days of *force majeure* so that the float is used up or to five days of *force majeure* so that the float is preserved? The resolution of this situation will depend upon how the provisions of the EPC contract have been written.

Additionally, the sponsor may want to circumscribe the ambit of the EPC contract's *force majeure* provision by restricting *force majeure* events to those occurring at or near the site and, in fairness to the EPC contractor, perhaps also to events occurring at or near the fabrication location of major or critical components (and possibly including events occurring on the transportation route of such equipment from the vendor's facility to the project construction site as well).

Logically, a party is excused by *force majeure* only to the extent that its performance is not possible. EPC contractors will usually attempt to replace the foregoing standard of "not possible" with a standard of "not commercially practicable under the circumstances" because, as noted above, most circumstances can be overcome if sufficient funds are expended, and this is probably a fair compromise. The EPC contract should require the affected party to give prompt notice of a *force majeure* event and submit a recovery plan and periodic reports while the event of *force majeure* or its consequences continue. Although affected parties who have not been punctual in complying with their *force majeure* notice obligations in an EPC contract often use various legal theories from "constructive notice" to "lack of prejudice" to "unconscionability" to defend themselves against the bar on claims for relief from *force majeure* events, if their notice of the *force majeure* has not been delivered by the deadline imposed under the EPC contract, U.S. courts have been fairly unforgiving in their enforcement of notice provision requirements in contracts.[2]

The parties should pay great attention to the *force majeure* provision. Once the EPC contractor begins work, it is likely to be the most read provision of the EPC contract. Rarely will the sponsor itself need to invoke the protection of the *force majeure* provision because the sponsor has relatively few obligations under the EPC contract other than to pay for the EPC contractor's work and payment is never excused by *force majeure*. More frequently, the sponsor will be trying to limit the purview of the *force majeure* provision so that the EPC contractor cannot claim *force majeure*. Any time the EPC contractor runs into difficulty or an unexpected problem and thereby faces a schedule delay or cost increase, it will attempt to use the *force majeure* provision as an escape hatch. Furthermore, if the sponsor has a contractual obligation to deliver the project's output or use to an offtaker or user by a certain date, which is generally the case, the sponsor should try to negotiate its agreement with its offtaker or user such that the offtake or use agreement contains an automatic schedule extension for the facility's target in-service date if the EPC contractor successfully claims *force majeure*. However, seldom will the offtaker or user agree to a *force majeure* provision in the offtake or use agreement that merely states that an event of *force majeure* under the EPC contract will automatically be deemed a *force majeure* event under the offtake or use agreement. It is even less likely that an offtaker or user will reimburse the sponsor for the sponsor's costs of a *force majeure* event involving the EPC contractor because the fear of cost overruns is usually one of the fundamental reasons why the offtaker or user has contracted with the sponsor to carry out the project.

If the sponsor is tenacious and lucky, the offtaker or user may agree that the offtake or use agreement will replicate the provisions of the EPC contract *verbatim* if the offtaker or user has not agreed that the offtake or use agreement can contain a provision that an event of *force majeure* under the EPC contract will be considered an event of *force majeure* under the offtake or use agreement (in which case the offtaker or user often requires an extension of the term of the offtake or use agreement that corresponds to the *force majeure* delay). Thus, the sponsor will be able to at least attempt to obtain the same treatment for the event under the terms of the offtake or use agreement as the treatment for the event that was obtained under the terms of the EPC contract. However, to make matters more complicated for the sponsor, often the offtake or use agreement will have been executed well before the EPC contract was executed, so that in reality all that the sponsor will be able to do is to try to pressure the EPC contractor into agreeing to a *force majeure* provision in an EPC contract that replicates the *force majeure* provision of the offtake or use agreement so that differing legal results under the contracts are less probable. Unfortunately for the sponsor, even if the *force majeure* provisions are identical in both contracts, different results are still possible because the question of whether or not *force majeure* has arisen may be before different arbitrators or judges (one case will be proceeding under the EPC contract and one case will be proceeding under the offtake or use agreement). In fact, often the offtake or use agreement and the EPC contract may even be governed by the law of different jurisdictions because the offtake or use agreement will usually be subject to the law of the jurisdiction in which the offtaker or user resides and the EPC contract will often be subject to the law of the jurisdiction in which the sponsor resides because the sponsor or its lenders may have more understanding of, or confidence in, the law of their home jurisdiction. Thus, different results are possible. The potential for inconsistent results and varying levels of relief will be of great concern to the sponsor's financiers, who will not want the sponsor to get caught in a trap in which the EPC contractor's performance to the sponsor has been excused but the sponsor's performance to its offtaker or user has not been excused. Hence, the sponsor could be at peril of a default under its offtake or use agreement with its offtaker or user if the facility does not enter service by the required deadline in the offtake or use agreement.

Several examples may be helpful to illustrate the nuances involved in *force majeure* provisions and claims.

Example 1. A state trooper stops and detains for two days a wide cargo carrier bound for the site because the state trooper believes that the truck is not being escorted properly. It turns out that the state trooper is wrong.

Example 2. A cargo ship carrying a steam turbine cannot dock for five days because no berth is available when it arrives in port.

Example 3. A vendor reserves a flat-bed railroad car to transport a gas turbine but the railroad company does not make the flat-bed available on time.

Example 4. A crane en route to the site cannot pass under a low bridge.

Example 5. A local traffic authority decides that a traffic light must be installed at the entrance to the construction site and the public road must be widened so a deacceleration traffic lane must be added to the public road.

Example 6. A subcontractor cannot find enough skilled pipefitters because another construction project in the area has added highly paid overtime shifts.

Example 7. A subcontractor goes bankrupt.

Example 8. A subcontractor's work on a foundation takes two weeks longer than expected because cement mixing trucks are in short supply.

Example 9. The overall, aggregate impact of various individual and unrelated items have delayed the EPC contractor's work.

Example 10. A fire marshal has halted construction work for an entire day to make an unannounced inspection of the construction site.

Example 11. A wetlands permit is revoked by the U.S. Army Corp. of Engineers (USACE) because it objects to the manner of the EPC contractor's dredging, although neither the USACE permit nor the EPC contract contains restrictions on methods that the EPC contractor can employ for dredging.

Example 12. An environmental group blocks the entrance to the worksite because it objects to the government's energy policy.

Example 13. The EPC contractor stops site clearing for four days because it discovers Indian arrowheads and charred animal bones and local authorities are called in to examine the remains. While the investigation is ongoing, a local farmer learns of the investigation and points out that the "artifacts" are just the remnants of a "fort" he and his brother built when they were children and played cowboys and Indians.

Example 14. The EPC contractor decides to take earth fill from a part of the construction site near a small river and gradually over the succeeding months the excavated areas become "marshy" as they fill with water from the river making that part of the site part of the waters of the United States and then the USACE informs the owner that a "wetlands" permit is required and all site work must cease.

Example 15. The government issues a terrorism alert and the EPC contractor withdraws its staff from the site but no event transpires.

At common law in the absence of a contractual *force majeure* provision in the EPC contract, none of the above cases is likely to be determined by a judge to be an instance of *force majeure* or impossibility.

Unfortunately, however, in all the cases above, if the EPC contract does contain a *force majeure* provision, whether or not *force majeure* can be successfully invoked by the EPC contractor will depend upon the specific language of the *force majeure* provision itself and what the judge or arbitrator believes was the intent of the sponsor and the EPC contractor when they entered into the EPC contract and wrote provisions of the *force majeure* clause.

Risk of Loss

When difficult circumstances arise, a contractual *force majeure* provision will determine if a party will be temporarily (or, in some rare cases, permanently) excused from its obligations. Who pays for costs that arise as a result of unforeseen circumstances is a different issue. Typically, the EPC contractor is given care, custody and control of both the construction site and work-in-progress no matter where it is located. Thus, the EPC contractor is responsible for safeguarding the site and the work and bears the risk of loss of the work. The EPC contract shall always provide that the EPC contractor is obligated to repair or replace any work that is damaged or lost for any reason whatsoever.

Usually, the EPC contract will provide that EPC contractor must repair and replace work that is damaged or destroyed at its own expense. In cases that are not *force majeure* and are not otherwise attributable to the negligent acts of the sponsor or any of the sponsor's contractors or agents, the EPC contractor will usually bear the cost of restoring lost work. In cases of *force majeure*, it is more common than not that the EPC contractor will still be responsible for these costs. Why? Because the EPC contractor is paid a premium in the EPC contract price to assume the risks inherent in construction regardless of when and why they materialize. Some EPC contracts, however, such as the International Federation of Consulting Engineers (FIDIC) forms, make a distinction between the risks that the sponsor must bear (usually called "owner's risks") and the risks the EPC contractor will bear. Usually risks allocated to the sponsor as "owner's risks" are risks that are uninsurable by the EPC contractor, such as war.

Builder's All Risk Insurance to Cover Risk of Loss

To protect itself against costs if a loss occurs, as will be discussed in more detail in Chapter 23, the EPC contractor (or sometimes the sponsor) will purchase builder's all risk (often referred to as "BAR" or "CAR" [contractor's all risk] insurance). When an incident of loss that is within the scope of coverage occurs, this insurance policy will cover the costs involved in restoring the loss, subject, of course, to any deductibles and limitations contained in the policy. Often, an EPC contractor that has a good relationship with the project's BAR insurer will itself advance repair costs from its own funds in anticipation of a claim being paid by the insurer so that the EPC contractor does not suffer a schedule delay or sacrifice valuable "float" time. This is usually one advantage of having the EPC contractor purchase the builder's all risk policy because the EPC contractor will be comfortable with the adjustment process of its own insurer and usually begin the rebuilding process before a claim has been awarded and thereby reduce the probability of delay of the project's in-service date.

Generally, protection under the builder's all risk insurance policy will be the "replacement value" of the work. However, certain events, such as war and the presence of nuclear radiation, are not covered by insurance. Frequently, therefore, the EPC contractor will limit its responsibility for reparation of lost work to the replacement of work destroyed by an incident that is within the scope of coverage of the builder's all risk insurance policy and other risks will remain the "owner's risk" (as discussed above). In this case, the sponsor should make sure to distinguish that the obligation of the EPC contractor to make reparations will be in force so long as the incident itself falls within the scope of the builder's all risk insurance policy, irrespective of whether or not the insurer actually makes a payment on the claim regarding the incident. This is necessary because there can be circumstances (as discussed below) in which an insurer could deny payment for a peril covered by the policy.

Occasionally, the sponsor will be able to obtain better rates for construction all risk insurance than the rates the EPC contractor itself can obtain. There may be several reasons for this. For example, the sponsor's long-standing corporate relationships with its own insurers or the complexity of the sponsor's project, that is, the facility that the EPC contractor is building may only be part of a larger facility being constructed and, therefore, the all-encompassing project construction policy may cover the facility that the EPC contractor is building. In these cases, EPC contractors usually will limit their responsibility for reparations to the actual proceeds of the sponsor's construction insurance policy as, and when, such proceeds are received by the EPC contractor. Thus, the sponsor will have to cover any shortfall or delay in receipt in the EPC contractor's proceeds if it wants the project to continue. Under this approach, in the case of an event that is not covered, or is disputed by an insurer, the EPC contractor will carry out its obligations to repair and replace work only if, and when, it receives the money to do so, either from the insurers or from the sponsor itself because the sponsor may determine that it is in its best interest to avoid delays and, therefore, may pay the EPC contractor to continue rather than wait to see whether its insurers will ultimately pay for the casualty in question.

In theory, one further complication could cause the insurance policy not to respond. The EPC contractor may have vitiated the policy's coverage, for instance, by making a misrepresentation to the insurance company in connection with the policy's issuance (such as incorrectly stating its accident or safety record, for example). Although this is a remote possibility, the EPC contract often provides that if this hypothetical case arises and the policy has been nullified, the EPC contractor will assume full reparation responsibility at its sole expense. Finally, if lenders are involved in a project, they will usually have negotiated the right to require insurance proceeds in excess of a certain threshold amount be paid directly to them to keep to repay their loans or for the lenders' subsequent disbursement of the proceeds to the EPC contractor if the lender determines to permit the sponsor to rebuild the project instead of keeping the insurance proceeds to satisfy their loans. Often EPC contractors overlook the issue that insurance proceeds may be held up or delayed but they remain liable to repair and replace lost work. Another area EPC contractors often fail to address is limiting their liability for damages to the sponsor's existing facilities in the case they are carrying out an expansion or upgrade of an existing facility in which they damage the existing facility. In that case the sponsor would make a claim under the sponsor's property insurance policy and then the insurer is likely to seek reimbursement from the insurance payout against the EPC contractor who caused the damage. For instance, a crane may have tipped over and damaged the existing facility.

Equitable Adjustments

Once the EPC contractor can demonstrate that an event of *force majeure* has occurred and has given the sponsor notification of the event in accordance with the terms of the EPC contract, most EPC contracts provide a mechanism by which the terms of the EPC contract can be adjusted to take into account the ramifications of the *force majeure* event. This adjustment is commonly referred to as an "equitable adjustment" or a "change." An equitable adjustment serves to vary one or more terms of the original contract to allow the EPC contractor to overcome the obstacles that have arisen as a result of the *force majeure* event. Most EPC contracts also delineate other events that are often not specifically included in the *force majeure* provision but will entitle the EPC contractor to relief if they occur. These events usually include changes in laws occurring after the date the EPC contract has been signed (as discussed in Chapter 9), delays caused by the sponsor or the sponsor's other contractors (see Chapter 10), or the discovery of hazardous materials or antiquities during construction (see Chapter 8). The EPC contract should specify that the EPC contractor will be entitled to an equitable adjustment only if the EPC contract expressly states that an equitable adjustment will be granted for the case in question. The EPC contract should state that under no circumstances will the EPC contractor be entitled to an equitable adjustment unless a circumstance is expressly contemplated by the equitable adjustment mechanism in the EPC contract. Ideally, these statements will preserve the fixed-price, turnkey philosophy that underpins the EPC contract (the tenet of which provides that the EPC contractor will bear all risks inherent in the work except those expressly assumed by the sponsor).

The EPC contractor should be required to provide the sponsor with immediate written notice if the EPC contractor believes that it is entitled to an equitable adjustment so that the sponsor can start assessing the event's consequences and the sponsor's alternatives in how to deal with it. The EPC contract should provide that if the EPC contractor fails to notify the sponsor that an event has occurred within some (short) period after the event's occurrence, the EPC contractor will not be entitled to an equitable adjustment in connection with the event. This will protect the sponsor from receiving "stale" claims for work that occurred in connection with events that the sponsor thought had been successfully completed, so that the sponsor had no reason to make an investigation of the incident when it could have done so effectively. Obviously, once work is complete and the workers that were involved have moved on to another project, it is more difficult to investigate the circumstances that preceded its completion.

In all the foregoing cases, the primary impact on the EPC contractor will be increased costs and/or delay. Thus, equitable adjustments essentially fall into two categories, the first is equitable adjustments for delay (equitable time adjustments) and the second is equitable adjustments for increased costs (equitable price adjustments).

Once the EPC contractor has demonstrated that a change in circumstances (such as a *force majeure* event or a change in law) that will entitle it to an equitable adjustment has occurred, it is imperative that the EPC contract require that the EPC contractor substantiate the increase in its costs and delay caused by the event in question. In essence, the EPC contractor should be compelled to prove the event's cost and consequences on the critical path of the baseline project schedule. Thus, the EPC contractor should be compelled to produce irrefutable evidence of its additional costs and delay. The EPC contract should require the EPC contractor to proceed with all work necessitated by the event even if the EPC contractor and the sponsor have not yet agreed upon the changes to the EPC contract that will be required. This will prevent further delays and costs from arising and ensure that the EPC contractor will not be able to coerce the sponsor into agreeing to the cost and schedule impact of the event just because the sponsor wants the EPC contractor to resume the work as quickly as possible so that the facility enters operations as soon as possible.

Equitable Price Adjustments

Many events can cause costs to increase, either because additional work becomes necessary, because work-in-progress must be postponed, or because work that has been damaged must be replaced. By definition, when an event giving rise to an equitable adjustment occurs, the parties to the EPC contract did not expect this event to occur. Cleary, this "modification" of the work was not in the original EPC contract price, which was based upon the work that the parties contemplated when they signed the EPC contract. When *force majeure* occurs, it is not the sponsor's or EPC contractor's fault that more or different work is involved. The question is, who will pay for the additional work. Typically, but not always, the sponsor will agree to bear the costs, most of which should be covered by insurance (as discussed above).[3] Often, however, the sponsor will agree to bear the cost only up to a certain limit or only to the extent that these additional costs can be passed on to the sponsor's offtaker and once these thresholds have been reached, the remaining costs will be borne (or shared) by the EPC contractor.

If the sponsor has agreed to bear all costs of a *force majeure* event, the sponsor's (and its financiers') preference will usually be certainty when it comes to such expenditures. Given the parties' objectives (the sponsor's desire to obtain a fixed price and the EPC contractor's desire to earn a reasonable profit on a new work), the sponsor and the EPC contractor generally will try to agree upon a lump-sum price so that the EPC contractor will bear the risk that the new work is more expensive than the parties estimated. In order to facilitate reaching agreement on a fixed price for the new work, the EPC contract should require the EPC contractor to deliver to the sponsor a comprehensive assessment of the costs of new work. This estimate should be itemized so that the sponsor will have a good point of reference for understanding all aspects of the additional work that the EPC contractor believes must be performed. The more detail the EPC contractor provides, the more confident the sponsor can be that the EPC contractor will not profit excessively from the need for the additional work.

EPC contractors are often reluctant at the time they sign the EPC contract to agree upon the content and form of the cost analysis that must be delivered to the sponsor if an "excused" event gives rise to additional work, but it is important to the sponsor that the EPC contract specify the level of cost detail that will be required if the EPC contractor makes an equitable price adjustment claim. Otherwise, valuable time may be wasted when the event arises and the sponsor must try to elicit the cost detail information that it desires from the EPC contractor, who may very well be busy worrying about other project issues (or even other projects). In extreme cases the sponsor may have to assemble this information itself, or use consultants, if the EPC contractor is not cooperative or responsive. For this reason the EPC contract should allot the EPC contractor a certain period during which it must assess the costs of this additional work and report its findings to the sponsor in writing. If the EPC contractor does not do this in the time allotted, the EPC contract should provide that the EPC contractor will not be entitled to cost relief from the sponsor and will itself be responsible for all costs of additional work.

The EPC contract should also always provide the sponsor with the option to direct the EPC contractor to proceed with the additional work on a "time and materials" basis if a lump-sum price cannot be agreed upon. This will allow the sponsor to continue to make progress on the project by paying the EPC contractor for the EPC contractor's out-of-pocket costs and labor expenses while a dispute is pending (which is why tables of labor rates and material costs should be included in the EPC contract). This time and materials approach may also give the sponsor some leverage to negotiate with the EPC contractor if the sponsor believes that the EPC contractor's fixed-price proposal is too high, either because the sponsor doubts that all the additional work that has been proposed is actually required or because the sponsor believes that the EPC contractor's price quote for the work is simply not competitive.

Equitable Time Adjustments

Once an event has caused a delay, two questions arise. The first question is "Can the delay be overcome?" The second is "Should the sponsor pay to have the work accelerated in order to preserve the originally planned in-service date for the project?" To answer these questions, the sponsor must understand the EPC contractor's situation at the time the delay occurs. Understanding the EPC contractor's progress at any point requires a brief digression into the organization of the project schedule.

A construction project is really an array of activities, some directly interdependent and some dependent only in the sense that they all need to be completed for the project to be finished. For example, the functional specification in the EPC contract may call for a separate machine shop building for storing tools and spare parts. Assume this building is prefabricated and could be erected in only three weeks at any point during the project's construction without interfering with the erection of the facility itself. In contrast to an activity like building the machine shop, a foundation must be set before the facility itself can be built. Suppose that, at the beginning of construction, heavy rain causes mudslides at the construction site and, as a result, the pouring of foundations is delayed. Obviously, the foundation for the machine shop building can be poured at a later date without much consequence. Unfortunately, the same is not true for the facility foundation. Timely pouring of the foundation for heavy equipment is obviously critical to preserving a project's schedule. As was explained in Chapter 11, pouring of the foundation is known as a "critical path schedule" item because its delay will delay completion of the project, whereas delay in pouring the foundation of the machine shop may not affect the project's scheduled completion date. The EPC contractor can easily "work around" the latter problem. Therefore, the machine shop foundation is not on the project's critical path schedule. Of course, any item can become a critical path item depending upon the amount of time it takes to be carried out and when the delay in question arises. Suppose in our example that the EPC contractor "worked around" the machine shop mudslide by delaying construction of the shop until one month before the project was scheduled to be completed. Then, at that time, heavy rain for three weeks again prevented framing of the machine shop building on top of the foundation. As a result of rain delaying the framing, the shop cannot be completed by the date that the project is scheduled to be completed. The machine shop construction will have become a critical path schedule item that prevents the EPC contractor from achieving the completion of the project by the targeted completion date. In this case, whether or not the EPC contractor is entitled to relief will depend upon how carefully the provisions of the EPC contract have been written.

The EPC contract should provide that only an event that, at the time it arises, affects activities that are on the critical path of the baseline project schedule will entitle the EPC contractor to additional time and solely to the extent that those critical path activities have been delayed as consequence of the event in question. A compromise to this unforgiving position of the sponsor would be to grant the EPC contractor a project schedule extension so long as the activity affected by the event in question is on the critical path of the project schedule at the time the event arises irrespective of whether or not the activity was originally on the critical path of the baseline project schedule.

Float

As explained in this chapter, a separate question is the duration of the period to accomplish a critical path activity and whether or not the EPC contractor has built "float" into activities on the baseline project schedule. For example, in the case of a power plant, suppose a foundation for the combustion turbine can be safely poured and set in 40 days but the baseline project schedule allowed for 50 days for this activity. If this activity is delayed and the EPC contractor must wait to pour the foundation, should the EPC contractor be entitled to a 40-day extension or 50-day extension? Again, this will depend upon whether or not the EPC contract has squarely addressed this issue of who "owns" any float that has been included as a buffer period in the baseline project schedule. Of course, it is entirely possible and even probable that a delay in pouring the combustion turbine foundation may well have a reverberating effect on other critical path activities and the EPC contractor may want (or need) an extension of even more than 50 days.

As a corollary to the above issue, sometimes an EPC contract will make expressly clear that the baseline project schedule allots to the EPC contractor a certain number of "buffer" days for its own use and that an event entitling the EPC contractor to an equitable time adjustment will not be permitted to cause a decrease in the number of buffer days allocated to the EPC contractor. Therefore, the EPC contractor will be entitled to an extension of the scheduled in-service date for the project in order to preserve the EPC contractor's buffer days that can be used by the EPC contractor in case delays occur that do not entitle the EPC contractor to equitable time adjustments (such as low labor productivity). This more lenient approach is consistent with typical practice in the construction industry.

While the EPC contract always sets a date (sometimes referred to by lawyers as a "date certain") for the EPC contractor's completion of the project, the EPC contractor is usually working toward a self-imposed target date that is earlier than the completion date set by the EPC contract. If the EPC contractor desires to preserve this buffer period in the equitable adjustment provisions of the EPC contract, it must make sure that the EPC contract has been properly drafted.

In summary, in making a claim for an equitable time adjustment, the EPC contractor should be required to explain which activities have been affected, which of the affected activities were on the critical path of the baseline project schedule, and which of the affected activities are now on the current critical path schedule as a result of the occurrence of the event. Next, the EPC contractor must be required to estimate the additional time, if any, necessary to complete the project as a result of critical path activities having been affected.

As a drafting and perhaps simply a psychological matter, the sponsor is probably best served not to use the term "equitable adjustment" because it seems to connote a measure of fairness and equity, which is not what the sponsor wants to elicit from a judge or arbitrator but rather the cold and antiseptic application of the EPC contract provisions and costs taking into account no other factors whatsoever—equity or otherwise. "Adjustment," not "equitable adjustment," is probably best used.

Change Orders

Once the sponsor and the EPC contractor have agreed upon modifications to the EPC contract made necessary by the occurrence of an event entitling the EPC contractor to an equitable adjustment, they will enter into a "change order." Although it sounds innocuous, a change order is actually an amendment to the EPC contract that memorializes the changes to the EPC contract that the parties have agreed upon in resolving the equitable adjustment. For example, in the case of the combustion turbine foundation mentioned above, the parties might have agreed to move the combustion turbine's location to a better elevation, delay its installation, and/or extend the project's completion date. All of the foregoing would be memorialized in the change order. In the case of the machine shop foundation, the parties may have simply agreed to a change order under which the EPC contractor would not be reimbursed for any costs that it incurred as a result of the postponement of the framing work but that the machine shop need not be completed until three weeks after the scheduled in-service date for the facility.

Most large projects will have many change orders. While change orders are generally for fairly simple and routine items and often signed on pre-printed forms, it is important for the sponsor to keep in mind that each change order is, in fact, an amendment to the EPC contract that alters and amends the EPC contract. As was discussed in Chapter 6, the sponsor's behavior, even before it has executed a change order, can result in a deemed change order if the EPC contractor has relied to its detriment on the sponsor's verbal instructions or actions.[4] For this reason, it is important that the intent behind the change order be stated in the change order itself. Generally, the intent will be that except as specifically amended by the change order, none of the other terms of the EPC contract are being amended and thus all terms of the EPC contract remain in effect. The sponsor must be wary that a change order executed in the field between construction managers has the potential to modify, or rescind, any provision of the EPC contract. A poorly composed change order has the potential to undermine the reason why the sponsor chose to enter into the EPC contract (paying a lump-sum price for facility completion by a certain date). On the other hand, there may be no need for a change order if the EPC contract makes clear that the work for which the EPC contractor is seeking additional compensation is already contemplated by the EPC contract. A party that offers additional compensation to a party that has failed to perform an obligation to which such non-performing party is already bound will not be liable to pay such additional compensation as promised because the non-performing party was already bound to perform the obligation and has given the promising party no additional consideration for its performance (that is, the non-performing party has assumed no additional responsibilities). This result serves to prevent extortionistic behavior. Of course, if the non-performing party actually offers any performance beyond the performance to which it was originally obligated, it will be able to collect the additional compensation promised. Generally, U.S. courts can look to see that there has been consideration given to the promising party for its promise to do something (otherwise it will be deemed a promise to give a gift [which cannot be legally enforced] and not a contract) but they will not look into the adequacy of the consideration involved because parties are free to set their own bargain—whether the price of the bargain is a million dollars or a sole peppercorn, either will be deemed "legal consideration" sufficient to create a contract.

Legal counsel for the sponsor should review all change orders before they are signed. It is also a good practice to have the EPC contract provide that change orders above a certain monetary threshold must be signed by a senior officer of the sponsor in order to be binding (although even this requirement can be waived inadvertently by the behavior of the sponsor's personnel or agents if the EPC contractor acts to its detriment in response to directions or promises by the sponsor's personnel or agents). In fact, if lenders have financed the project, often the loan agreements for the project will require that the sponsor obtain the lenders' prior written approval before either entering into any change order that is in excess of a specified amount, waives requirements of any performance test or postpones the project's completion date.

If the sponsor has borrowed money to finance construction of its facility, it should pay careful attention to the terms of its loan agreement regarding change orders. Lenders will typically restrict the sponsor's ability to enter into change orders above a certain monetary threshold or concerning the project's schedule or capability without the lenders' prior written consent. The sponsor should make sure that the threshold required for the lenders' consent is high enough so that simple and minor changes can be implemented without seeking lenders' consent. Lenders' consent may require the affirmative vote of two-thirds, three-quarters or even 100 percent of the lenders that are in the loan syndicate, depending upon the issue involved. A consent from lenders may also take weeks or months to obtain and therefore may be an expensive proposition for the sponsor. While the sponsor can attempt to negotiate a time limit on the lenders' consent action in the loan agreement, lenders will probably not agree to this type of provision. Lenders may also charge a fee for their consent as well (especially if, in their view, the risk profile for their loans will change as a result of the proposed change order) and usually must also be reimbursed for the out-of-pocket expense they incur (such as legal and engineering fees) in connection with their evaluating proposed change orders. During the consent solicitation process, if the EPC contractor cannot work around the problem that is the subject of the change order, work may be halted and delay costs may be mounting while the sponsor awaits its lenders' approval. A good example of how this type of situation might arise could be the EPC contractor's encountering an underground plume of oil for which the sponsor, in conjunction with local authorities, has developed a remediation strategy (such as simply capping the plume with concrete and then re-configuring the footprint of the facility) but which strategy cannot be implemented without lenders' consent, which may take months for the lenders to analyze the proposed strategy.

Often, the sponsor's lenders will even enter into an agreement between themselves and the EPC contractor (see Chapters 13 and 14) that will require that the EPC contractor notify and/or obtain the lenders' written consent before the EPC contractor signs any change orders over agreed-upon value and project schedule changes of or in excess of a certain duration

The EPC contract should also provide that once an item has been addressed in a change order, the EPC contractor may not again raise a claim with respect to such item unless a new event that would entitle the EPC contractor to an equitable adjustment under the EPC contract has occurred. This type of preclusion will compel the EPC contractor to make a responsible and exhaustive assessment of the issues that have arisen in connection with the occurrence of each event. The EPC contract also should bar the EPC contractor from making any claims for an equitable adjustment once the sponsor has agreed that the facility is complete and released the final payment (or a large portion of the final payment) to the EPC contractor in order to avoid surprises and maintain financial leverage over the EPC contractor.

Directed Changes

The signing of the EPC contract memorializes the commercial arrangement between the sponsor and the EPC contractor and also "freezes" the description of the facility that the EPC contractor will design and build. However, the sponsor's expectations about its facility often continue to evolve after the EPC contract has been executed. For this reason, sponsors usually reserve the right to make unilateral changes to the facility's design and construction so long as the sponsor pays the EPC contractor for the additional costs involved. If this is the case, the sponsor and the EPC contractor usually will negotiate a change to the EPC contract to cover this unanticipated work. However, if they cannot agree, the EPC contract will usually provide that the sponsor can direct the EPC contractor to perform additional work (often referred to as a "directed change" or "owner directed change" or sometimes, unfortunately, simply a "change order") on a "cost plus a profit margin" basis. This often raises many issues that the EPC contract should also address, such as whether this additional work will be paid for in advance, on a percentage-completion basis as the work is completed, or only once the work is complete. The EPC contract also should address whether or not the sponsor, in connection with ordering additional work, has the power to alter the EPC contractor's promises (the performance guarantees discussed in

Chapter 17) relating to the output and efficiency of the facility. The additional work may entail significant design work by the EPC contractor and change the entire risk profile of the EPC contractor's undertaking. In terms of facility performance, the EPC contractor will usually have "over-designed" the facility to make certain the facility will meet the facility's performance requirements promised in the EPC contract. Consequently, a change in the sponsor's requirements for performance of the facility may serve to reduce or eliminate the EPC contractor's "over design" safety margin entirely and, therefore, be of great concern to the EPC contractor.

It is typical for the EPC contractor to prevent material changes in the project's risk profile by insisting that the EPC contract provide that the sponsor will be able to require the EPC contractor to carry out additional work only so long as such additional work remains below a certain dollar value or remains consistent with the scope of the facility's specifications. For instance, the EPC contractor would probably not want to add an additional train to an LNG facility but may have no problem adding an extra tank for LNG storage capacity. The EPC contractor will not be required to carry out any work in excess of such limit unless the EPC contractor agrees upon the terms of such work and the changes to the EPC contract required by this additional work.

Constructive Changes

Although a well-drafted EPC contract will forbid the EPC contractor from requesting an equitable adjustment for a circumstance unless the EPC contract expressly states that the occurrence of such circumstance will entitle the EPC contractor to an equitable adjustment, EPC contractors can still (occasionally successfully) seek equitable adjustments based upon the sponsor's behavior (both before and during the execution of the work).

A "constructive change" can arise when the sponsor has acted in a manner that is inconsistent with the terms of the EPC contract. Therefore, the EPC contractor must perform its work in a more costly or time-consuming manner than could have reasonably been anticipated by the EPC contractor when it executed the EPC contract. "Constructive" change can occur when the sponsor:

(a) requires work not called for by the EPC contract;
(b) supplies inadequate design documents, materials or equipment;
(c) requires work be conducted in a more expensive manner than necessary;
(d) applies higher quality standards than are mandated in the EPC contract;
(e) forces the EPC contractor to comply with an erroneous interpretation of the EPC contract;
(f) denies a justified time extension, which has the consequence of requiring the EPC contractor to accelerate the work unnecessarily; or
(g) requires the EPC contractor to perform the work out of the EPC contractor's planned sequence.[5]

A prerequisite for the EPC contractor's recovery under any of the above events is that the sponsor must have acted in contravention of the terms of the EPC contract. Thus, if the EPC contract authorizes the sponsor to take any of the above actions, they cannot be the basis for a constructive change claim by the EPC contractor. It is possible for the sponsor to mitigate its risk of a constructive change claim from the EPC contractor by setting up a dispute resolution mechanism in the EPC contract that can serve to resolve differences of opinion regarding the types of issues listed above. As will be examined in Chapter 24, the sponsor's engineer, independent experts and dispute resolution boards can all be employed to resolve differences of opinion between the sponsor and the EPC contractor. The conclusions of these professionals generally will bar an EPC contractor's constructive change claims unless the EPC contractor can clearly demonstrate that the professional resolving the dispute acted with a malicious intent to cause the EPC contractor harm.

Cardinal Changes and Contractual Abandonment

In some cases, even though the sponsor may have the unilateral right to direct the EPC contractor to make changes, the EPC contractor may still be able to claim that the sponsor's changes were so great or fundamental that the sponsor has made a so-called "cardinal" change in the work or even abandoned the object of the EPC contract entirely by asking the contractor to undertake work that is materially different from that which was the object of the EPC contract.[6] If the sponsor fails to adhere to the intent and object of the EPC contract (another example of why the intent clause is critical), the EPC contractor may be entitled to collect its actual damages from the sponsor. On the other hand, claims for lost profits in U.S. courts will usually be denied because lost profits are generally too speculative to be awarded (common law principles dictate that damages must be estimable and also foreseeable by the defendant to be awarded) and not mere hypothesis and conjecture.

Cumulative Impact Changes

Occasionally, an EPC contractor may argue that the overall impact of many individual change orders has created a situation whose aggregate effect entitles the EPC contractor to an equitable adjustment. While the EPC contractor may be able to prepare a convincing case for a claim of this nature, a simple statement in the EPC contract that such "cumulative impact" claims cannot be made by the EPC contractor should preclude this type of claim from being successful.

Notes

1 See *Seitz v. Mark-O-Lite Sign Contractors, Inc.*, 210 N.J. Super. 646 (App. Div. 1986), in which a court held that it was not *force majeure* or impossibility when a contractor's only expert sheet metal worker became ill and could not perform his work because the worker's diabetic condition (requiring an amputation) was not a "force" beyond the control of the contractor.
2 See *McDevitt Street Co. v. Marriott Corp.*, 713 F. Supp. 906 (E.D. Va. 1989), enforcing a seven-day notice provision. Also see *Milford Power Co., LLC v. Alstom Power, Inc.*, 822 A.2d 196 (Conn. 2003), holding that the sponsor's knowledge of an incident giving rise to *force majeure* could not allow an EPC contractor to overcome a notice requirement for the submission of an equitable adjustment claim because the sponsor could not have known that the EPC contractor planned to lodge an equitable adjustment claim. And see *Feeney v. Bardsley*, 66 N.J.L. 239, 49 A. 443 (1901), ruling that a duty to provide notice must be observed. Also see *Omi Specialties-Washington, Inc. v. Esprit De Corp.*, 1989 WL 34326 (D.D.C. March 28,

1989), holding that informal means of notice do not satisfy a formal notice requirement contained in a contract.

3 See *A. Kaplen & Sons v. Housing Authority of Passaic*, 42 N.J. Super. 230, 232 (App. Div. 1956), denying the contractor's claim for monetary relief because the contract specifically barred monetary claims for delays of any nature.

4 See *Headley v. Cavileer*, 82 N.J.L. 635, 82 A. 908 (1912), noting that even clauses that require amendments to a contract to be in writing can be superseded by an oral agreement between the parties.

5 See *Pellerin Construction, Inc. v. WITCO Corporation*, 169 F. Supp. 2d 568 (E.D. La. 2001), denying a contractor's claims for cardinal change and work out of sequence because the contract put the contractor on notice of such potential changes.

6 See *Empire Foundation Corp. v. Town of Greece*, 31 N.Y.S.2d 424 (Sup. Ct. 1941), ruling that a significant change in the routing of a sewer line by the owner entitled the contractor to relief. Also see *C. Norman Peterson Co. v. Container Corp.*, 218 Cal. Rptr. 592 (Ct. App. 1985), holding that change orders and extra work ordered by the owner were of such a magnitude that the owner had abandoned the contract and the contractor could, in order to prevent unjust enrichment, recover the value of its work under the legal theory of *quantum meruit*, which allows a party to recover the value of materials and work that it has supplied to another party in the belief that it would be compensated for the value of the materials and work furnished. In order to sustain a claim for *quantum meruit* against receipt of work or materials, a supplier must have a reasonable expectation that the recipient would compensate the supplier for the materials. Thus, under a claim for *quantum meruit*, a subcontractor to an EPC contractor will not be successful in recovering its costs from the owner if the supplier is not paid by the EPC contractor (absent a state statute or provision in the EPC contract otherwise) because the owner has not agreed to pay the subcontractor and the subcontractor could not have had a reasonable expectation that the owner would pay the subcontractor. And see *Insulation Contracting & Supply v. Kravco, Inc.*, 209 N.J. Super. 367 (App. Div. 1986), holding that an owner was not liable to a subcontractor of the owner's general contract even though the owner received the benefit of the subcontractor's work because there was no contractual relation (known as "privity of contract") between the owner and the subcontractor.

Chapter 13

The EPC Contractor's Failure to Perform

The Concept of Default

A party's failure to perform its obligations under a contract is usually referred to as a "breach of" or "default under" the contract. Once a breach or default is demonstrated to a court, the non-breaching party can seek remedies from the court for its injuries. However, it can sometimes be difficult to determine exactly when a party is in breach of a contract, especially in the context of cases involving "minor" and "immaterial" deviations from contractual terms. Therefore, in an attempt to avoid the need to demonstrate to a judge that a breach has taken place, it has become common practice for parties to list which provisions of the contract will be considered by the parties a breach of the contract and automatically lead to a default under the contract if a party fails to comply with any of them. It is also fairly common to include certain events that may not be within the control of the parties in the list of contractual defaults. One such event, for example, might be the loss of a particular permit by a party. Upon the occurrence (and usually continuance) of any of the events listed as defaults in the contract, the affected party can proceed directly to request a remedy from the court rather than be caught in the intermediate step of trying to prove that its counterparty's non-compliance was substantial enough to constitute a breach of the contract.

Remedies at Law vs. Remedies in Equity

Typically, common law and most U.S. state statutes require courts to award monetary damages to a plaintiff unless the plaintiff can demonstrate that the payment of monetary damages will not be adequate to compensate the plaintiff for the harm that it will suffer from the breach of the contract. In cases in which the plaintiff can demonstrate that there is no adequate remedy "at law" (monetary damages), the court has the power to award so-called "equitable relief" pursuant to which the court can direct the actions or behavior of the defendant, rather than just require the defendant to pay monetary damages to compensate the plaintiff.[1] Most equitable relief today probably takes the form of temporary restraining orders or injunctions that prohibit a party from undertaking or continuing an action. In rare cases, however, such as personal services contracts (contracts for performers, for example), a court may actually use its equitable powers to force a party to perform an obligation. While most jurisdictions in the United States hold that engineering and construction contracts are not personal service contracts and, therefore, an EPC contractor will typically be required only to pay damages to an injured owner rather than perform its work, it is conceivable that an EPC contractor could be forced to perform a contract and build a facility.[2]

Adequate Assurance

In most jurisdictions, a party to a contract is entitled to seek "adequate assurance" if it has a reasonable anticipation that its counterparty intends to breach its contract. Thus, in the case of "anticipatory breach," if a party can demonstrate that it has reasonable grounds to suspect that its counterparty cannot or will not perform, the suspicious party is entitled to require its counterparty to demonstrate that it will perform. If the defending party cannot provide an adequate assurance of its ability to perform, the court may order a termination of the contract.[3] What the defending party must demonstrate in order to prove that it will perform its obligations under the contract will be a "question of fact" and will depend upon the circumstances of the particular case. The remedy of adequate assurance gives affected parties the ability to take counter-measures and mitigate their damages rather than just wait for an upcoming breach and then take remedial and possibly also legal action.

The principle of mitigation of damages is one of the central tenets of common law. It requires that an injured party must take reasonable measures to mitigate its damages. Although this principle may seem useful, it may not be so helpful for complex contractual relationships like an EPC contract. It may be difficult in an adequate assurance claim to convince a court that a counterparty is unwilling or unable to perform its obligations and, even if the court is convinced, the court may not be inclined to order the counterparty to perform if monetary relief would be adequate compensation for the injury to be sustained. Therefore, the aggrieved party (usually the owner) will still be left to seek damages and the services of another EPC contractor and therefore the owner will find itself in the same position that it was in before it sought adequate assurance. Thus an adequate assurance suit can often be a costly waste of time.

The Owner's Remedies

Basically, as discussed above, an owner has two basic remedies available to it when it seeks redress against an EPC contractor in a court. Either the owner can seek monetary damages if it believes damages will adequately compensate it for the harm that it has suffered or the owner can seek an equitable remedy, such as specific performance requiring the EPC contractor to perform (or injunctive relief prohibiting the EPC contractor from taking a certain action) if the owner believes that it has, or will be, irreparably harmed by the EPC contractor's behavior.

Generally, the EPC contract will enumerate a list of remedies available to the owner if the EPC contractor does not perform an obligation under the EPC contract. Inclusion of such a list will make it very difficult for the EPC contractor to dispute the owner's entitlement to avail itself of any of these remedies. The most typical remedy is what is commonly referred to as a "self-help" because it will provide that if the EPC contractor does not perform an obligation under the EPC contract, the owner will have the right to perform the work on the EPC contractor's behalf, or engage someone to perform the EPC contractor's work, and charge the EPC contractor any costs in excess of those the owner would have paid to the EPC contractor. In actuality, this is the most typical remedy available to the owner in a court action. Therefore, this remedy probably does not need to be enumerated in the EPC contract but including it will help avoid disputes about whether the parties intended that the owner pursue another remedy before or in lieu of pursuing the right to step in and complete the work. In practice, however, the "self-help" remedy is usually not very practical because the party that is truly in the best position to do the work is the EPC contractor and not the owner or another EPC contractor. Another EPC contractor becoming involved in the project will usually create additional expense and delay because the substitute EPC contractor will need to familiarize itself with the project and assess the state of the work thus far completed and this is therefore highly undesirable for all project participants. In fact, as the EPC contractor knows, the "self-help" remedy is generally viewed by owners and lenders as the remedy of last resort and will be used by an owner or its lenders only if no other practical remedy exists or all other practical remedies have been exhausted. As will be seen later in this chapter, this vicissitude often gives the EPC contractor substantial leverage in disputes.

Exclusivity of Remedies

Often, EPC contractors will request that the EPC contract contain a provision stating that the remedies expressly noted in the EPC contract will be the owner's sole and exclusive remedies for the EPC contractor's breach of the EPC contract regardless of whether other remedies may be available to the owner "at law" or "in equity." The owner should resist such a limitation on its remedies because the owner should not contractually relinquish other remedies (such as the right to "adequate assurance" discussed above) that may be useful in certain cases. Furthermore, the owner's counsel may not have evaluated all the remedies available to the owner under the law that governs the EPC contract and, therefore, the owner might be waiving rights that may be helpful (or even necessary) in the future if the EPC contractor fails to perform in accordance with the EPC contract. After all, the EPC contractor's performance is completely within its own control. If the EPC contractor is concerned about the owner's panoply of remedies under law, the EPC contractor should simply not breach the EPC contract. While all this rhetoric sounds convincing, once the conference room banter subsides and it is time to get a deal signed, often a compromise is reached. From the owner's point of view it is usually fair to provide that if the EPC contract itself provides for an explicit remedy for a particular breach (such as liquidated damages for failure to attain a required facility performance level), such remedy will be the exclusive remedy for that breach no matter whether or not other remedies are available to the owner at law or in equity. It is the subject of debate whether a party can waive its right to "equitable" remedies. Equitable remedies are remedies granted by the court in its discretion under its equitable powers, so a waiver of these remedies may, luckily, not be enforceable because private citizens may not have the right to "exorcise" courts of their equitable powers. However, in cases in which the EPC contract does not specifically attach a remedy to a default, the EPC contract should make clear that the owner will maintain its right to seek any remedy available to it "under the law."

Engineering Difficulties

As was noted in Chapters 7 and 11, the most serious mistakes in a project can occur during its engineering and design phase. An error in design that is not discovered until construction is completed can usually not be fixed without modification or destruction of some of the facility (generally an expensive and time-consuming proposition). In the case of a power plant, for example, suppose (as mentioned in Chapter 7) that the rail unloading and handling system for a two-unit coal-fired plant has not been adequately designed to feed sufficient fuel to both boilers at the same time. Therefore, the plant cannot achieve its target electrical output. A problem like this cannot be corrected without re-design and re-construction. In the above example, suppose that the turning radius of a coal supply train has been miscalculated by the EPC contractor and the owner purchased the site based on the miscalculated radius. What problems might the owner have when it needs to buy additional property? First, the landholder will probably realize that it may have "holdup" value now in the additional parcel that the owner needs. Second, what if the additional property that the owner needs has a factory operating on it? Or, what if the owner's facility abuts a body of water rather than land? As will be seen in Chapter 16, the EPC contractor should be required to test the entire facility rather than just its individual units if so-called "common facilities" are shared between a facility's individual units or trains. In practice, as a result of the magnitude and uncertainty involved in large-scale rectifications, owners and EPC contractors generally reach a liquidated monetary settlement to compensate the owner for lost revenues and/or additional costs arising from these situations rather than implementing a physical solution.

Shortcomings in design can lead to constructability and reliability issues as well. For example, in cases of refurbishments and upgrades of existing facilities, engineers who do not have extensive experience in capital improvement projects may not design layouts that take into account the difficulties involved in construction in confined areas such as existing buildings and underground vaults. Projects like these are particularly complicated if some portion of an existing facility must continue to operate while refurbishments to other parts must be made. Disruption is usually of paramount concern in cases of refineries, processing facilities or combined heat and power plants (CHPs) because their owners will want to minimize any curtailment of steam, hot water and electricity service to the district or campus served by the CHP. In these cases it is imperative for the owner to specify in the EPC contract that construction activities must be carried out in a manner that will minimize impacts on existing service. The owner may even want to delineate periods during which interruptions in service will not be permitted such as during class sessions or examination periods.

Another design concern for owners can be reliability. Suppose a natural gas-fired power plant has been designed with a compressor on the inlet pipeline because the existing pipeline pressure on the local natural gas distribution company's (LDC's) network is not sufficient. One natural gas compressor might be the most economical design for the owner's power plant in terms of cost and energy usage (that is, the energy used to power the compressor). However, when the compressor undergoes routine maintenance or experiences an equipment failure, the power plant will be subject to a complete outage because it will have no natural gas supply. During this outage, the owner probably will not be able to collect revenue because the plant will not be capable of generating energy. Worse yet, the owner may be required to pay liquidated (or even actual) damages to the plant's offtaker for the plant's unavailability if the compressor has failed unexpectedly.[4] It may make more sense for the owner to install two compressors each rated at 50 percent of the size of the originally intended compressor (even though this option might initially be more costly than a single compressor). This solution might even require more space and decrease the plant's efficiency by increasing the plant's own energy consumption, usually referred to as "parasitic" or "house" or "station" load, because two compressors will be running instead of one. All these factors must be balanced against the advantage that the owner will still be able to operate the power plant, albeit at a reduced output if one of the plant's compressors fails or requires maintenance. Many similar examples (such as pumps, tanks, pipes, piers, etc.) exist of this balancing of operational philosophy against the capital investing program. The owner should play an active role in understanding these issues. Avoiding design phase problems and optimizing the cost/reliability balance requires proactive vigilance and owners will often enlist the help of an owner's engineer in this regard (see Chapter 11).

Without question, the owner should have the unfettered right to review all design documents other than (as was discussed in Chapter 11) design documents and shop drawings relating to internal workings of components that the EPC contractor will purchase from vendors (such as computer software and hardware, transformers and generators). Practically speaking, it is probably not realistic for the owner to review all design documentation. Therefore, the EPC contractor and the owner will usually agree upon a list of design documentation to be delivered to the owner for its review. However, the EPC contract should make clear that the owner reserves the right to review all design documentation regarding the project, whether or not it is included on the design document "deliverable" list.

As was mentioned in Chapter 11, as a legal matter, the owner should "review" and not "approve" design documentation. Of course, if necessary, the owner should have the right to comment on design documentation and the EPC contract should require the EPC contractor to address and resolve all of the owner's comments to the owner's satisfaction. Usually, time periods are allotted for the owner's review of design documents it desires to study so that the EPC contractor's performance of the work will not be delayed while awaiting the owner's comments. Clearly, however, if an issue that the owner believes is important cannot be resolved, the owner should take whatever action under the EPC contract (that is, mediation or arbitration) or at law that is necessary to prevent the facility's improper construction. In this case, while the owner may be at risk for a delay claim by the EPC contractor if the owner delays the EPC contractor's work while it tries to resolve the issue, it is usually better to deal with a problem as early in the design phase as possible before the EPC contractor has a chance to compound the problem by basing further design and construction on the deficient design that the owner has identified.

Construction Difficulties

Construction difficulties can be divided into two categories. The first, and most benign, category is problems created by low productivity and delay. The second, and more serious, category is poor or defective workmanship. Neither of the cases is easy to rectify unless the owner has a good relationship with the EPC contractor.

Labor that is inefficient or inexperienced typically creates costly delays as activity schedules become misaligned and, consequently, equipment and workers sit idle while they wait for predecessor tasks to be completed. If the equitable adjustment and change order provisions in the EPC contract have been properly written, the EPC contractor will be responsible for the costs of these delays. As was mentioned in Chapter 12, irrespective of the date that the EPC contractor has promised the owner for the completion of the work, the EPC contractor will most likely be targeting a completion date earlier than the date that has been guaranteed by the EPC contractor for completion of the work (as will be discussed in detail in Chapter 17). While this target date is not binding on the EPC contractor, it is generally wise for the owner to know this target date (and the EPC contractor will usually be willing to share it with the owner) so that the owner can see if the EPC contractor is behind the EPC contractor's own "internal" schedule (as opposed to the baseline project schedule attached to the EPC contract).

If the EPC contractor does not complete the work on time, as will be seen in Chapter 17, the EPC contract will generally provide that the EPC contractor will pay liquidated damages to the owner for the delay in completion. While these damages often do not compensate the owner for the full amount of the owner's actual losses as a result of the delay, the liquidated damages amounts are usually high enough so that the EPC contractor will not want them to continue to accrue and will finish the project as soon as possible in order to minimize these payments. In most cases these liquidated payments are capped between 15 and 30 percent of the EPC contract price in order to protect the EPC contractor in cases of serious delays in its completion of the work.

A protracted delay can become a serious problem for the owner in several regards. First, if the owner has borrowed money to finance its project, construction loans usually will have to be repaid to lenders if the facility is not placed into commercial operation by an agreed-upon date, which will be specified in the loan agreement. The construction lenders' commitment to "convert" their construction loans into long-term loans by extending their maturity will expire if the facility is not in service by the agreed-upon date contained in the loan agreement. Expiration of the commitment to extend long-term credit to the owner on a "date certain" is premised upon the rationale that if the owner cannot successfully complete construction by an agreed-upon "drop dead" date, the construction lenders should be able to begin to take action against the owner. If they choose to foreclose on the uncompleted project and sell it in a public auction in an attempt to satisfy their debt, this often will wipe out the owner's equity investment in the project because the value of an uncompleted project is likely to be less than the loans lent for its construction. This foreclosure may not help the construction lenders very much either because they are likely to be in the same position as the owner was vis-à-vis the tardy EPC contractor. If, instead of a foreclosure sale, the construction lenders accede to the EPC contract in the place of the owner (because they will have negotiated this accession or "step in" right in a separate "direct" agreement [as discussed in Chapter 12 and will be discussed in Chapter 14]) they will not have any remedies in addition to those remedies that the owner had against the EPC contractor. Thus, the construction lenders really have three options if their loans are not repaid. First, they can sell the project (for what is likely to be a huge loss) in a foreclosure sale. Second, they can take over the project themselves and deal with the EPC contractor directly. Third, they can keep the owner "in place" and let the owner continue to deal with the EPC contractor. This last approach is the most typical and usually construction lenders will try to extract a fee from the owner (or sometimes increased interest rates) if they take this approach and permit the owner to maintain its ownership interest in its project.

Another complication resulting from an extended delay can occur if the owner has negotiated a commitment from an offtaker or user to purchase or use the plant's output or capacity. Like the construction lenders' commitment to convert their construction loans into long-term loans, the offtaker's or user's offtake or use commitment will usually expire if the facility is not commercially operable by an agreed-upon deadline. Hence, the owner could lose its contracted revenue source if this deadline in the offtake or use agreement passes without the facility being put into service and the offtaker or user decides to terminate the offtake or use agreement. Clearly, this can be a disaster. If the owner has entered into an offtake or use agreement, usually both the owner and its lenders are depending upon the offtake or use contract for return of their investment. If the offtake or use contract is terminated by the offtaker or user, it will probably be a bankruptcy judge who will decide the fate of the project. (See "early operation" in Chapter 16 for protections that can be included in the EPC contract to mitigate this "meltdown" risk).

In instances of construction delays and reduced construction productivity, the owner's most effective defense is careful monitoring of the EPC contractor's activity in order to pre-empt or truncate delays before they become consequential. If the owner believes the EPC contractor is falling behind schedule, it should notify the EPC contractor immediately in writing. The EPC contractor may respond that the project is on schedule or temporarily lagging but that the project will be completed by the agreed-upon "guaranteed" completion date. Essentially, a "trust us, we are the experts" retort. The owner should not rely unquestioningly on the EPC contractor. It should ask the EPC contractor to provide a detailed schedule to substantiate its "everything is o.k." response. If the owner is not satisfied with the EPC contractor's response, the owner should engage a reputable industry scheduling expert to study the state of the project and assess the project's current schedule. The owner should, subject to its legal counsel's advice on the attorney–client privilege in the relevant jurisdiction, share the results with the EPC contractor. This legal advice is necessary because the owner will not want to waive any rights that it might have to keep its own studies confidential and protected from a court's power to "discover" and review these records if they are not covered by the attorney/client privilege (as was discussed in Chapter 4). The EPC contractor may actually benefit from the industry expert's information. Even if the EPC contractor ignores the information, the information's transmission to the EPC contractor can be used to establish a record for later proceedings against the EPC contractor should the owner need to pursue them. Should the owner continue to have concerns about the EPC contractor's performance after sharing the information, it is in the owner's best interest to send a letter to the contractor noting that, as outlined in the "intent clause" of the EPC contract, "time is of the essence" and the owner has placed great trust and confidence in the EPC contractor's commitment to finish the facility by the agreed-upon date, and in reliance upon this intent, and the EPC contractor's representations in the EPC contract that the EPC contractor would place the facility in service by the scheduled in-service date, the owner has placed a huge sum of its own and its lenders' capital at risk.

While the EPC contract will purportedly "cap" the damages that the EPC contractor must pay for delay at a liquidated amount, as was alluded to above and will be discussed in Chapter 20, it may actually be possible for the owner to convince a court to disregard this limitation on damages if the EPC contractor has willfully or recklessly breached its obligations to finish the work on schedule. Although it will be a difficult burden of proof for an owner to overcome in order to demonstrate that an EPC contractor has acted in "bad faith" or with "recklessness" in failing to acknowledge or correct a schedule or productivity problem, letters informing the EPC contractor of the grave effects that a long delay will have on the owner's investment may later allow the owner to argue in court or an arbitration that the EPC contractor was aware of the extensive losses the EPC contractor would cause as a result of an extended delay.

Cases of shoddy workmanship are usually relatively easy for the owner to handle unless the consequences do not exhibit themselves until after the warranty period of the EPC contract expires. For example, in the case of a power plant, what if a part, such as the steam turbine thruster bearing (which absorbs the horizontal thrust of the rotator shaft), heats up to a temperature above that which is typical for such bearings in similar steam turbines but is still below the temperature rating for the alloy of which the bearing is made? In cases like that, in which no "defect" is apparent or likely to appear in the near future (that is, before the warranty period expires), the owner's best recourse is (again) to commission an expert study to analyze the long-term ramifications of the "deviation" in question and determine whether or not the EPC contractor has adhered to the intent of the parties by building a highly reliable facility. Logically, disputes can arise if the functional specification does not contain enough specificity (that is, not specifying a continuous concrete pour without cold joints in the case of the turbine foundation mentioned in Chapter 7 and not specifying a maximum permitted operating temperature in the case of the thruster bearing). Without a doubt, all situations cannot be addressed in the functional specification or the EPC contract, which is why the owner's first and last resort will often rest in the "intent" clause of the EPC contract.

Financial Difficulties

An EPC contractor's financial difficulties can paralyze a project. If an EPC contractor does not stay current on its expenses (such as payroll and invoices from subcontractors and vendors), work ceases. Unless the owner is willing to advance these payments, there is little that can be done. The longer the hiatus, the more expensive will be the cost to reinstate the work and, therefore, the owner must quickly determine the nature of the EPC contractor's financial problem.

The owner must assess whether the EPC contractor itself is insolvent (or headed in that direction) or whether the EPC contractor has just lost interest in the owner's project because problems (such as overruns and delays) have already wiped out any profit that the EPC contractor expected to earn on the project. In the latter case, to avoid expensive equitable adjustment claims and further delays, the best strategy may be to give the EPC contractor some form of bonus or incentive for completing the work early in order to try to rekindle interest in the project. Should the EPC contractor be headed for bankruptcy, the most efficacious plan may be to negotiate a settlement so that the EPC contractor will withdraw from the project and the owner can replace the EPC contractor with another EPC contractor. Once an organization starts the slide toward bankruptcy, its employees generally turn their attention away from their work and toward finding new jobs. In addition, the velocity of the organization's slide usually accelerates as its banks terminate its credit lines. If the EPC contractor will not agree to a settlement for its exit from the owner's project, the owner can try to terminate the EPC contract based upon the EPC contractor's default.

Complex commercial contracts, including EPC contracts, usually contain specific clauses that provide that a default will be deemed to have occurred if a party files for bankruptcy protection or has a bankruptcy claim filed against it that it is not able to dismiss reasonably promptly. However, these clauses originated before there was a Federal Bankruptcy Code in the United States and they are probably unenforceable in the United States because they contradict the legislative intent behind the U.S. Federal Bankruptcy Code (which is to provide assistance in the resuscitation of a debtor's finances). Under the U.S. Federal Bankruptcy Code, a party cannot terminate a contract with the EPC contractor solely because the EPC contractor has sought the "protection" of the U.S. Federal Bankruptcy Code, regardless of what the contract states. In addition, a bankruptcy filing gives rise to a broad injunction known as the "automatic stay" that prevents creditors and contract counterparties from exercising rights that arose before the filing against the bankrupt company. An owner, for example, cannot terminate an agreement based on a pre-bankruptcy filing breach once the EPC contractor has filed for bankruptcy. The owner must rely upon one of the other events of default contained in the EPC contract (and that default must have occurred after the bankruptcy filing) in trying to terminate the EPC contract (such as a default for failure of the EPC contractor to pay its subcontractors on a timely basis or default for the EPC contractor's failure to prosecute the work in a timely fashion if these have been included in the EPC contract). If the bankrupt EPC contractor has not "tripped" one of the other defaults contained in the EPC contract, the owner could still try to claim that the EPC contractor's slow and uncertain performance violates the "time is of the essence" understanding that the parties had in entering into the EPC contract. The strength of this claim will depend upon the existence of a "time is of the essence" clause and relative strength of the "intent" clause contained in the EPC contract as well as whether or not the bankruptcy judge is convinced that the EPC contractor will not perform. The owner must act as quickly as possible when financial difficulties loom.

U.S. Bankruptcy Example

Suppose that an EPC contractor that is a Delaware limited liability company (EPCco) sets up two subsidiaries that are also Delaware limited liability companies. One subsidiary will be based in New York and build projects in New York (NYsub) and the

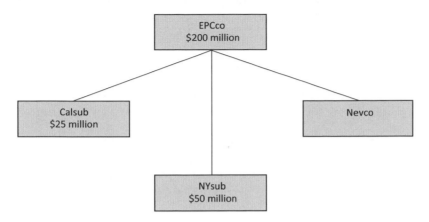

Figure 13.1 Hypothetical EPC contractor.

other subsidiary will be based in California and build projects in California (Calsub). Suppose also that EPCco borrows $100 million from public investors in the form of debentures in order to fund EPCco's general operations. Suppose that Calsub borrows $25 million from banks in California (and pledges its own Calsub assets in California) to fund its activities in California, while NYsub does not need to borrow any money to fund its activities. Also suppose EPCco has assets of $200 million and NYsub and Calsub have assets of $50 million and $25 million, respectively (Figure 13.1). In this case, the consolidated financial statements of EPCco will show assets of $275 million.

If an owner is entering into an EPC contract with NYsub, is there $275 million worth of assets "behind" this EPC contract? No, just the $50 million of assets of NYsub. If the owner obtains a parent guaranty from EPCco, then there will be $250 million of assets "backing" the EPC contract (assuming the $100 million of debenture proceeds have been used to purchase services and meet payroll, etc.). Not $275 million of assets. Why? Because the assets of Calsub are beyond the reach of NYsub and maybe even beyond the reach of EPCco since these assets are pledged to Calsub's lenders.

Calsub could file for bankruptcy and not legally affect the operations or standing of either its parent EPCco or its affiliate NYsub (although, as a practical matter, the credit and investment market might view its credit "profile" or "rating" differently). Of course, the result of Calsub's bankruptcy could be that EPCco's equity investment in Calsub is wiped out and its membership interest in Calsub must be turned over to Calsub's lenders in satisfaction of Calsub's debt to them, but, otherwise, if EPCco has issued no guaranties on behalf of Calsub, EPCco will remain unscathed as a result of Calsub's bankruptcy (so long as it has not fraudulently expropriated Calsub's assets for its own benefit).

In fact, a bankruptcy of EPCco itself might not affect its subsidiaries Calsub or NYsub if these subsidiaries can operate without EPCco's assistance (unless, of course, EPCco was a guarantor of their debt, in which case their loan agreements are likely to provide that the bankruptcy of their guarantor, EPCco, will cause an event of default under, or even accelerate, their loans). How is this scenario possible? Suppose EPCco issues $75 million of additional public debentures to fund EPCco's acquisition of a smaller EPC contractor in Nevada called Nevco. Now suppose Calsub loses money on several projects in California (perhaps as a result of overruns that are not reimbursed by owners as equitable adjustments under the relevant EPC contracts). As a result, Calsub is not as profitable as expected and pays no dividends to EPCco but instead requires a cash infusion from EPCco so that Calsub can continue to operate. Now, EPCco may be in the situation in which it cannot pay its own debenture holders because of the unexpected cash "drain" of Calsub. If the problem is truly serious and not just temporary, EPCco may decide to file for Chapter 11 protection under Chapter 11 of the U.S. Federal Bankruptcy Code in order to try to reorganize its debts with its creditors (primarily the debenture holders) or "convert" its debt into equity, in which case all or perhaps just certain classes of its original equity holders will be "wiped out" and lose their investment. In fact, if this "cash crunch" is not indicative of a systemic problem, EPCco may have an easier time obtaining credit from lenders once it is operating under Chapter 11 protection. The U.S. Federal Bankruptcy Code encourages the extension of credit to debtors and grants special rights to creditors who have extended credit to a debtor after it has filed a petition for protection under Chapter 11 of the U.S. Federal Bankruptcy Code. Sometimes, even a primary position in the collateral of other lien holders will be given to new lenders who make loans that will help preserve the viability of a debtor's business. These are commonly referred to as "DIP" or "debtor-in-possession" credit facilities and certain banks specialize in this business of lending money to debtors.

Procedural and Substantive Consolidation of Cases

If EPCco files a petition for Chapter 11 protection, its subsidiaries NYsub, Calsub and Nevco may or may not choose to file their own petitions for Chapter 11 protection. (All legal entities must file their own petition in order to seek protection under the U.S. Federal Bankruptcy Code.) In the above example, it is possible that only EPCco and Calsub would file bankruptcy petitions. Each separate petition will create a separate bankruptcy case. For convenience of administration, however, a bankruptcy court may choose (for reasons of administration) to consolidate separate cases of affiliates into one bankruptcy proceeding. This practice of consolidation of cases for administrative reasons is known as "procedural consolidation." In the case of "procedural consolidation," the "estate" of each debtor (such as EPCco and Calsub) remains separate and creditors must deal with the entity with whom they originally contracted even though there is only one proceeding. In certain cases, debtors may request, or creditors may require, the assets and liabilities of multiple entities to be pooled for purposes of making distributions to claims holders. This is known as "substantive consolidation" and is rare in cases where the debtor and its primary creditors do not agree to implement it.

Rejection of Contracts

If either of the above consolidation events occurs in the example above of Calsub's and EPCco's "bankruptcy," creditors or counterparties of NYsub or Nevco probably need not be concerned if these companies can operate without EPCco's financial or tendered assistance. However, for an owner that has entered into an EPC contract with Nevco or NYsub and relied upon a parent guaranty of its EPC contract from EPCco, there can be one important and distressing exception. In bankruptcy, EPCco will have an opportunity to "cherry pick" which of its contracts it will perform and which of its contracts it will "reject" and not perform. This right of selection applies only to the debtor's so-called "executory" contracts (that is, those contracts under which performance has not yet been discharged by both parties). For instance, a guaranty is not "executory" and not "rejectable."

If a debtor "rejects" a contract, the counterparty to the contract can file a claim for breach of contract and other damages against the debtor. Unless the counterparty has taken collateral for the debtor's obligations under the "rejected" contract, however, its claim will be an unsecured, pre-petition (that is, a claim arising before the debtor's bankruptcy filing), which will likely be subject to the same payment reduction to be suffered by all the other debtor's "pre-petition" creditors once a plan of reorganization is approved for the debtor. If an owner has accepted a guaranty of Nevco's or NYsub's performance from EPCco and this guaranty is not honored by EPCco, the owner may have big problems. If the owner's lawyers are doing their job, they will have given the owner some possible leverage to prevent EPCco from seeking bankruptcy protection. An EPC contract between an owner and Nevco or NYsub should contain, as one of the EPC contractor's events of default, the bankruptcy of EPCco or the failure to honor the parent guaranty made by EPCco. This event of default will give the owner the right to terminate the EPC contract and thus the EPC contractor and (indirectly EPCco) can expect to lose any profit it (or they) hoped to earn because the owner may retain a replacement EPC contractor to complete the project and the completion cost charged by this replacement contractor is likely to be higher than the remaining balance of EPC contract price, and, thus, the excess will be charged to the EPC contractor. Although it is questionable whether or not this type of provision can really be enforced and in practice using it as a threat will depend upon whether there is actually any profit to be made on the EPC contract (because the EPC contract might have already had overrun or delay problems that have wiped out its expected profit), its presence may cause the EPC contractor's parent to think twice about making a bankruptcy filing in order to facilitate a reorganization because, in the process, the profit on the EPC contract may be jeopardized by the bankruptcy filing if the owner does, in fact, avail itself of the bankruptcy default clause's purported protection.

Of course, if the parent entity of the EPC contractor is really in severe financial condition, a bankruptcy filing may be unavoidable because any three creditors can typically start the process to push a debtor into bankruptcy by filing a so-called "involuntary" petition against the debtor for its reorganization or liquidation. In fact, as a company approaches the so-called "zone of insolvency," its directors and managers must begin to act with the interests of its creditors in mind and not solely the interests of its owners (as is usually the case in corporate law), so they may feel that they no longer have the discretion to refrain from making a bankruptcy petition filing (although in states such as Delaware such consideration of creditors is only necessary once the debtor is insolvent and not just in the "zone of insolvency"). In addition to the fact that the owner will no longer have the parent company's "deep pocket" to look to for performance of the EPC contract if the EPC contractor does not perform, the bankruptcy filing by EPCco will almost certainly be an event of default under the owner's loan agreement with its lenders if the owner has financed its project. As opposed to the situation mentioned above that noted that a party's own bankruptcy cannot cause a default under one of its own agreements, the bankruptcy of another party is enforceable as an event of default in a non-bankrupt party's agreement. The lenders will have included the bankruptcy of EPCco as an event of default under their loan agreement because EPCco was the "credit" behind the EPC contractor itself, whose assets may not have been extensive enough to make the lenders confident that Calsub could complete the project if financial or technical difficulties arose. In this case, if the owner's lawyers have properly negotiated the loan agreement, the owner will have a period in which to try to obtain a "replacement" guarantee of the EPC contractor's performance from a guarantor that is acceptable to the lenders. Of course, the owner's finding a "replacement" guarantor is more easily said than done because it is unlikely that a party that is unaffiliated with the EPC contractor (even a surety, for instance) will have any interest in stepping up to the obligations of an EPC contractor whose parent went bankrupt. Sometimes in these situations, it is the owner, or the owner's own parent, that will guarantee the EPC contractor's performance to the lenders in order to prevent the owner's own investment in the project from being wiped out in a foreclosure by the lenders.

The Slide Toward Insolvency

Once a company recognizes the possibility of bankruptcy, a new array of advisors and experts may emerge and begin to control its destiny and guide its business strategy. The owner may have new personalities, issues and agendas to deal with and should consider hiring its own insolvency professionals. Once a company enters the trajectory toward bankruptcy, it becomes difficult for the company to fight the inertia of its falling market capitalization (if it is a public company) and the gravity of unpaid accounts payable. While directors of a solvent enterprise have a fiduciary duty to act in the best interest of a company's owners, and creditors must rely solely on their contractual rights, directors of an insolvent company must take the interests of the company's creditors into account because they are the company's primary stakeholders (shareholders have no economic interest in an insolvent company). In many of these cases, the owner should realize that the ailing company's financial course cannot be reversed without a bankruptcy proceeding. The owner should start making plans to contain the economic damage from the bankruptcy filing rather than pretending that the situation will work itself out. In fact, if one EPC contractor slides into bankruptcy during the project, the trustee or creditors of the EPC contractor could conceivably seek to "claw back" (as a preferential payment) any payments such as liquidated damages made by the EPC contractor to the owner if such payments were paid within the 90 days prior to the EPC contractor's bankruptcy filing (or 365 in the case where the owner is an affiliate of the EPC contractor) if the payments were made in respect of antecedent debt and without fair value being received by the EPC contractor therefor. If the owner remains diligent in billing and collecting it may preserve the "ordinary course of business" preference defense in order to avoid a preference claim against it.

Subcontractor Difficulties

Detecting and handling subcontractor problems can be difficult. A well-structured EPC contract will require the EPC contractor to notify the owner of any significant disputes that arise between the EPC contractor and any of its subcontractors. It should also require the EPC contractor to give notice, or even seek the owner's prior consent, before it amends or waives any material provision of any significant subcontract. Furthermore, each time the EPC contractor requests payment from the owner, the EPC contractor should be required to certify that the EPC contractor is not in arrears with any of its subcontractors. These provisions are intended to provide signals to alert the owner to problems. An irascible or incompetent subcontractor can disrupt a project. A subcontractor's irresponsible performance, or failure to perform, can endanger the work and the schedule while at the same time creating an expensive legal nuisance for the owner. In many jurisdictions a subcontractor that has not been paid for its work can place a lien on the materials it has incorporated in a facility. This may become a problem for the owner, even where the owner has no contract with the subcontractor and is not responsible for the EPC contractor's debt. The subcontractor may be entitled to proceed against the portion of its work that was incorporated into the facility and seek its return or sale to satisfy its claim against the EPC contractor. While the EPC contract should require the EPC contractor to solicit the agreement of its subcontractors to waive these rights to file "materialmen's" or "workmen's" liens, in the United States waivers of this sort are usually invalidated by statutes or determined by courts to be void because they are against public policy.[5] The EPC contract should require that the EPC contractor defend the owner's property against any liens filed by any subcontractor (either by posting a cash bond or by indemnifying [reimbursing] the owner for any costs or losses that the owner incurs or suffers as the result of the lien).

The EPC contractor and the owner sometimes have divergent interests when it comes to problems with subcontractors. If the owner has financed its project, the owner's loan agreement with its lenders will usually provide that the owner must keep its facility free of liens. Sometimes, liens are permitted so long as they do not exceed a certain dollar amount in recognition of the fact that minor disputes often arise on a large infrastructure project and small lien amounts will not affect the value of the facility. If an owner fails to keep its facility lien-free as required by the lenders, it may face a default under its loan agreement and possible acceleration of the maturity of all its loans. The facility is the lenders' collateral for their loans and they do not want any other parties such as subcontractors to have any rights in their collateral. In fact, as was discussed in Chapter 2, while lenders will have a "first" mortgage or "first" lien on the owner's facility to secure their loans to the owner, many materialmen's lien statutes permit materialmen's liens to take priority over all other liens including mortgage liens. Thus, lenders' collateral may be at risk to materialmen's liens. Even if the first priority position of the lenders' lien is not at risk, lenders generally do not like to share their collateral with any other party (whether *pari passu* with the lenders' interest in the collateral or junior interest). Many state statutes provide that a senior lienor must act reasonably to preserve and not adversely affect the value of collateral of a junior lien holder. In the event of trouble, lenders want free rein to do as they please with their collateral without regard to the interests of another lienor. Lenders also do not want to have to attorn to another lienor for any "questionable" actions that they may have taken (such as not paying to rotate the combustion turbine rotor at a power plant that has been temporarily "mothballed" as a result of the owner's bankruptcy). That is why lenders rarely lend in situations where another lienor will have an interest in "their" collateral.

An owner may simply want a lien removed to avoid a potential default under its loan agreements even if the EPC contractor has to pay more to the subcontractor than was bargained for in the original subcontract because the owner will not bear this expense. The EPC contractor is likely to have a different view. If the EPC contractor believes that it has a *bona fide* (valid) claim against the subcontractor that filed the lien, the EPC contractor is likely to lose much of its leverage in the dispute if it pays this subcontractor. Additionally, the EPC contractor may have an ongoing relationship with the subcontractor on other projects and may be reluctant (or eager) to engage with the subcontractor, while the owner may not have other dealings with the subcontractor.

Without a doubt, the owner will not have much interest in watching the jockeying between the EPC contractor and its discontented subcontractor while deadlines for the in-service date of its facility under its loan and offtake or use agreements are approaching. If necessary, the owner can resort to withholding payment and/or exercise the owner's right to remove any person or entity from the project if such person or entity is not capable of performing the work assigned or whose presence poses a hazard to the work. This may be a sub-optimal course of action since the EPC contractor could lodge an equitable adjustment claim for delay against an owner if the EPC contractor is required to "eject" a subcontractor from the project and spend time and money searching for a replacement subcontractor. Nevertheless, it can be used to send a signal to the EPC contractor that the owner will not tolerate situations that unnecessarily jeopardize the in-service date of the project.

The Owner's Options

If the courses of action outlined above do not accomplish the desired result, the owner still has the ability (at its own expense) to direct a change in the work (as was discussed in Chapter 12) or suspend or terminate all or any part of the work (as will be discussed in Chapter 15). Although it appears the owner has myriad options available when it needs to solve any one of the problems described above, the truth is, unfortunately, that all are expensive stopgap measures that may sometimes prevent a project's economics from further deterioration but that are unlikely to reverse any erosion of the project's economics that result from the difficulties discussed above. Agile and effective deployment of these options is a poor substitute for the tactic that can prevent almost all these difficulties from arising—choosing an experienced and responsible EPC contractor. Experience comes at a price as the cliché goes. Seldom, if ever, will the most responsible EPC contractor for a project put forth the lowest bid for the project. An experienced EPC contractor is generally better at estimating the time and effort involved in completing work. The owner should bear this in mind when comparing bids. Low bids may signal that an EPC contractor has not properly estimated the cost of a project, or it may simply be that the EPC contractor is eager to win the business, with a desire to break into a new market or segment (maybe EPC work itself).

Perhaps one of the best ways to assess the EPC contractor's commitment to a project before executing an EPC contract has nothing to do with financial or technical analysis. It has to do with personalities and people. The owner should request a meeting with the EPC contractor's most senior management before signing the EPC contract to assess their commitment to the project and their willingness to dedicate organizational resources to the project throughout its duration—from preliminary engineering through to warranty repair work. A project is not simply an assembly of steel and concrete, it is an economic enterprise. Therefore, quantitative, qualitative and even psychological analysis is compulsory when choosing an EPC contractor.

Notes

1 Courts of equity in England in the early part of the first millennium were created because common law judges recognized relatively few causes for legal actions (and also relatively few remedies). A complainant or respondent that believed it had been dealt an unjust verdict could turn only to the king (acting through his chancellor) for relief. Perhaps because chancellors were often bishops and had a penchant for presiding over atonement, they would often order breaching parties to honor their promises rather than merely require the breaching party to pay monetary damages for the harm it had caused (as would be the remedy typically imposed by a common law judge). Eventually, by the later part of the first millennium, separate courts of equity generally had been merged with common law courts (although their remedies have survived and expressions such as "sitting in equity" or "equitable powers" are still common in judicial opinions). See Dawson, John P., Harvey, William Burnett, Henderson, Stanley D., *Cases and Comment on Contracts* 37 (4th ed., The Foundation Press, Inc., 1982).

2 See *Sithe Fore River Development, LLC et al. v. Raytheon Company*, Supreme Court, New York County, Index No. 601206/01, in which an owner sought an injunction requiring specific performance as the remedy to enforce the terms of a guarantee issued by a parent of an EPC contractor that had abandoned the construction work. Also see *Sokoloff v. Harriman Estates Development Corp.*, 96 N.Y.2d 409 (2001), denying an architect's claim that there was an adequate monetary remedy for its failing to deliver its plans to the plaintiff and noting that specific performance is a proper remedy "where the subject matter of the particular contract is unique and has no established market value."

3 Section 2–609 of the New York Uniform Commercial Code provides:

   (1) A contract for sale imposes an obligation on each party that the other's expectation of receiving due performance will not be impaired. When reasonable grounds for insecurity arise with respect to the performance of either party the other may in writing demand adequate assurance of due performance and until he receives such assurance may if commercially reasonable suspend any performance for which he has not already received the agreed return.

   (2) Between merchants the reasonableness of grounds for insecurity and the adequacy of any assurance offered shall be determined according to commercial standards.

   (3) Acceptance of any improper delivery or payment does not prejudice the aggrieved party's right to demand adequate assurance of future performance.

   (4) After receipt of a justified demand failure to provide within a reasonable time not exceeding thirty days such assurance of due performance as is adequate under the circumstances of the particular case is a repudiation of the contract.

4 Under offtake or use agreements, owners are typically required to pay liquidated damages if their facility is not available for use as planned. For instance, in the case of a natural gas-fired power plant, an owner may agree that its plant will be available 98 percent of the time during summer peak electricity demand hours (usually 7 a.m. to 9 p.m.) and 90 percent of the time on an annual basis. For example, if the owner has agreed that its plant will be available for operation for 8,000 of the 8,760 hours in a year and it turns out the plant was in fact available for service for 8,000 of the 8,760 hours of the year, the plant would be said to have 100 percent "reliability" (8,000 ÷ 8,000) and 91 percent "availability" (8,000 ÷ 8,760). In the case that it was only available for service for 7,000 hours during the 8,760 hours in the year, it would be said that the plant would have a "reliability" of 88 percent (7,000 ÷ 8,000) and an "availability" of 80 percent (7,000 ÷ 8,760) and, therefore, it would generally have to pay some amount of liquidated damages (which are usually paid according to a sliding scale) for its poor performance. Sometimes, however, instead of liquidated damages, offtake agreements can require an owner to supply a replacement for the offtake during periods when its plant is unexpectedly unavailable (or reimburse the offtaker for the offtaker's costs of purchasing replacement offtake to be used in the offtaker's processing [if the offtaker is a refinery, for instance] or the offtaker's customers [if the offtaker is a steam utility, for instance]). In either case, whether the owner must set up diesel generators or package boilers or purchase electricity on the spot market to replace its offtake, this usually is an expensive penalty for the owner to absorb so it is advantageous for the owner to agree upon realistic availability targets in its offtake or use contract. The owner can usually determine a reasonable availability target from other owners' experiences because, in the case of tested technology such as combustion gas turbines, literally, millions of operating hours of experience may have already been logged on a particular gas turbine model and made available for review by vendors to their potential customers.

5 See *Wm. R. Clarke Corporation v. Safeco Insurance Company of America*, 15 Cal. 4th 882, 938 P.2d 372 (1997), ruling that public policy prevented the waiver of a mechanic's lien.

Chapter 14

The Owner's Failure to Discharge Its Responsibilities

In comparison to the EPC contractor, as was explained in Chapter 10, the owner will have relatively few obligations under the EPC contract. The owner's primary obligation will be to make payments to the EPC contractor as the EPC contractor completes different stages of the work, generally referred to as "milestones." Other obligations of the owner usually relate to furnishing permits and interconnections for utilities, such as fuel and telecommunications, on a timely basis so that the EPC contractor's work is not delayed. Often, however, the owner's payment of additional money to the EPC contractor can compensate the EPC contractor for any delays that increase the EPC contractor's costs, such as re-mobilization costs for idled craft labor and rented equipment. In the case of long delays, however, the owner may not be able to keep the EPC contractor economically whole, to the extent that the EPC contractor incurs an opportunity cost in tying up its own personnel who could be expending their efforts and energy on one of the EPC contractor's other projects. This often becomes an issue if the EPC contractor is using its own engineering manpower because these personnel could be assigned to another project instead of being held idle so they can be available promptly if the owner's delays cease. Generally, the EPC contractor will not have the same problem with construction labor crews. They will usually be employees of one of the EPC contractor's subcontractors. (EPC contractors usually do not have large contingents of trade labor but often have field supervisors on their own payroll who rotate from project to project for the EPC contractor to supervise subcontractors' crews and work.)

Recognizing that the impact of delay of the EPC contractor's work can be fairly easily quantified and rectified, it is to the owner's advantage to set up the EPC contract so that any delays or disruptions caused by the owner will entitle the EPC contractor to an equitable adjustment but no other remedy (such as damages). Thus, a change order granting the EPC contractor an extension of the schedule and compensating the EPC contractor for its incremental costs would be given by the owner but no default or breach of the EPC contract by the owner would be deemed to have occurred as the result of a delay caused by the owner. Naturally, the EPC contractor should be entitled to stop work if the owner does not stay current on its payments of the EPC contract price. It is reasonable to categorize the owner's failure to make timely payment as a default under the EPC contract. Consequently, so long as the owner is current in payments to the EPC contractor, any additional costs that the EPC contractor incurs as a result of delays caused by the owner can be compensated equitably by means of an equitable adjustment and, thus, there need be no owner default triggered under the EPC contract's provisions.

It is also in the owner's interest not to allow the EPC contractor to include default provisions in the EPC contract relating to the solvency of the owner. Default provisions that are triggered by the bankruptcy of the owner are not a problem (because, as explained in Chapter 13, *ipso facto* [meaning occurring simply as a result of something] provisions are probably unenforceable under the U.S. Federal Bankruptcy Code). Provisions that create defaults for the owner's failure to pay its debts as they become due can be problematic. The owner should resist any efforts by the EPC contractor to include such events as defaults if the EPC contractor is receiving payment from the owner (especially if they are worded broadly such as "admits its inability to pay its debts as they become due" as opposed to "receives a going concern exception from its auditors in their report"). If the EPC contractor is able to terminate the EPC contract for the owner's insolvency even though the EPC contractor is being paid, the owner's cost of replacing the EPC contractor is likely to be very high and probably uneconomic from a profitability standpoint. The owner's loss of the EPC contract will likely be tantamount to the owner's losing its ability to make any profit on its project.

The chances are good that the owner's equity investment will be wiped out if the EPC contractor is able to terminate the EPC contract. Furthermore, if it does not require a sizable further loan from them, the owner's lenders may very well lend the owner funds or take over the project so that payments can continue under the EPC contract because if the EPC contract is terminated and the plant is never built, or becomes significantly more expensive to build, the lenders are unlikely to ever be repaid in full. The owner must guard against being an economic casualty in a fight between the EPC contractor and the owner's lenders. How could this happen? It can happen if the owner has cash flow problems (perhaps, as a result of costly change orders). Thus, the owner might be facing termination of the EPC contract by the EPC contractor as a result of a payment default under the EPC contract. If this is the case, the owner, if it has financed its project, will probably also be in default under its loan agreement with its lenders, who will likely have listed the occurrence of a default under the EPC contract by the owner as a default by the owner under their loan agreement. If the lenders vote to declare a default under their loan agreement they can begin the process of seizing their collateral, which will include the EPC contract itself. In either case, the lenders themselves will usually have the right to use their own funds to cure the owner's default under the EPC contract by paying the EPC contractor directly (although lenders have the option, it is rarely really used in practice because lenders will often have trouble agreeing among themselves on whether or not and how much money to continue to loan to the owner and also have trouble agreeing upon the magnitude of the borrower's problems [that is, whether the problems are temporary or endemic]). In contemplation of this very eventuality, at the time the lenders make a loan to the owner they will customarily enter into a so-called "direct" or "consent to collateral assignment" agreement between themselves and the EPC contractor giving them the right to cure any defaults by the owner under the EPC contract. If the lenders decide to seize their collateral (the project and its contracts) from the owner, this seizure might be done by way of a foreclosure (essentially an auction conducted under court supervision) or by means of a negotiated settlement between the owner and its lenders. In either case, it is likely that the owner's equity investment in the project will be lost.

These games of brinkmanship with the owner's equity investment will not be played if the EPC contract does not contain a default provision triggered by the creditworthiness of the owner but instead merely incorporates the concept that the EPC contractor can stop work if it does not receive payment from the owner as due (after the relevant grace period, if any, given to the owner under the EPC contract has expired). If the EPC contractor has ceased work for an agreed-upon period as a result of non-payment by the owner, the EPC contractor should, in fact, be entitled to terminate the EPC contract.

Although no one will be eager to hypothesize about financial crisis scenarios during the negotiation of the EPC contract, the owner should make all reasonable attempts during negotiations to limit its contractual defaults under the EPC contract to only one—the situation in which the EPC contractor is not being paid and, therefore, is suffering serious economic harm that makes it financial suicide (and just plain stupid) for the EPC contractor to continue to work. In any other case, the EPC contractor should be indifferent as to whether or not the owner is paying its other creditors, or even operating under the protection of the U.S. Federal Bankruptcy Code (unless a court-ordered liquidation of the owner is imminent). Furthermore, if lenders have lent money to the owner, the EPC contractor should take comfort that it will be paid by the owner or its lenders because lenders will have little chance of their loans being repaid unless the project is constructed. A project site, equipment, stockpiles of rebar, bags of concrete and a file cabinet of expired permits and contracts are likely to be worth only a small fraction of the principal amount of (and accrued interest on) the lenders' loans. At times, lenders' lawyers may argue that provisions of mortgage and security agreements are sacrosanct and lenders will not consider making any change to these time-tested provisions because lenders are in the business of foreclosing on assets in order to maximize and preserve asset value when necessary. In reality, foreclosures are not a profitable line of business for banks. Anything the owner can do to minimize the likelihood of a foreclosure is probably worth a few extra minutes of discussion at the negotiating table, even though raising the notion of the project encountering financial trouble can be taboo before lenders have committed their funds (because the owner will not want lenders to change their minds about lending to the project).

"No Damages for Delay" Provisions

Customarily in construction contracts, one sees a "no damages for the owner's delay" clause. This provision makes clear that the contractor will not be entitled to seek any damages from the owner for the contractor's increased costs or lost profits if the owner delays the contractor's progress. However, the contractor almost always will be entitled to additional time to the extent that it can no longer perform the work in the time allotted under the contract. This type of provision has developed because delays in complicated construction projects are almost inevitable. Owners do not want to assume responsibility for the costs associated with these delays—even if the delays are caused by the owner. U.S. courts have typically upheld these provisions.[1]

Unlike traditional construction contracts, EPC contracts usually contain an express provision that is the antithesis of the "no damages for delay." EPC contracts generally provide that the EPC contractor will be automatically entitled to any additional costs it incurs as a result of a delay caused by the owner. Most owners acquiesce to these terms. If this provision is acceptable to the owner, the owner should ensure that the scope of this provision is not unnecessarily broad. If its scope is not circumscribed appropriately, the very reason that the owner entered into an EPC contract—to shift the risk of design and construction contingencies to the EPC contractor—could be compromised. The owner should only be responsible for delays caused by it and its agents and contractors. The EPC contractor must be required to demonstrate that the source of the delay was the owner or its agents or contractors. A provision that merely makes the owner responsible for delays "not attributable to the EPC contractor" or "not attributable to the EPC contractor's breach of its duties under the EPC contract" is dangerous. Such broadness will create a *terra incognita* between delays for which the parties intended the EPC contractor to be responsible and those for which the parties intended the owner to be responsible. These broad "attribution" clauses should be avoided. They will create a question of fact (the intention of the parties) to be decided by an arbitrator, judge or jury after evidence has been presented by the parties rather than just a question of contractual interpretation (or simply reading comprehension) to be resolved by a judge or arbitrator as a matter of law without any presentation of evidence being necessary.

Although the EPC contractor's typical position that the owner should be responsible for any delays in the work caused by the owner sounds logical, historically, in construction contracts owners are usually not responsible for the costs of delay caused by the actions of the owner or its other contractors.[2] Thus, the owner may want to provide expressly in the EPC contract or even a separate agreement between all contractors at the site that the EPC contractor must coordinate with other contractors of the owner at the EPC contractor's sole cost and risk. Hence, the EPC contractor should not be free to make optimistic assumptions about how and when other contractors will complete their work and then have recourse to the owner if these irresponsible assumptions are incorrect. Misparked trucks, crowded laydown areas, ships that overstay their berthing window, insufficient work camp facilities and rail siding congestion are all examples of simple problems that can compound, interfere and use up valuable schedule float. For instance, an EPC contractor should probably not be permitted to assume in its project schedule that the owner's contractors will encounter no delays in the completion of their work. Instead, a well-drafted provision will try to demarcate a clear boundary between delays for which the EPC contractor will be held responsible because they are foreseeable and those for which the EPC contractor will not be held responsible because the delays could not have been reasonably contemplated by the EPC contractor in its original evaluation of the work.

Active Interference and Loss of Efficiency

Regardless of what is or is not written in the EPC contract, an owner's "active interference" in the EPC contractor's work will expose the owner to liability to the EPC contractor. Aside from purely belligerent behavior on the part of the owner, even the owner's benign interdiction in helping to resolve the EPC contractor's confusion or disorganization can form the basis for a successful "active interference" claim by the EPC contractor. As a precaution, if the owner must become involved in a design or construction matter with the EPC contractor for any reason, the owner should be mindful to note in written communications with the EPC contractor that the owner's intention is to assist the EPC contractor in its interpretation of the EPC contract's terms and not to interfere with the EPC contractor's performance of the work. This may help establish a record of the intent of the owner that could be difficult for the EPC contractor to refute if the matter becomes the subject of a formal dispute proceeding.[3]

As a corollary to an "active interference" claim, a contractor might try to make a "loss of efficiency" claim predicated upon the assertion that the owner's actions or requirements have resulted in the EPC contractor's loss of the efficiency or productivity that it anticipated when it entered into the EPC contract. If the owner actually does require the EPC contractor to alter its work methods or practices (for example, preventing blasting even though the owner originally said blasting would be permitted), it may be hard for the owner to defend itself against this type of claim without a written statement from the EPC contractor acknowledging (implicitly or explicitly) that the EPC contractor will not be impacted by such a change. In the case that the owner does ask the EPC contractor to make a change, the owner can probably best protect itself by asking the EPC contractor to confirm in writing that the requested change will not increase the EPC contractor's costs (or, if it will increase the EPC contractor's costs, asking the EPC contractor to quantify for the owner the cost of the change requested).[4]

Notes

1   See *Wells Brothers Co. v. United States*, 254 U.S. 83 (1920), in which it was held that a contractor was not entitled to seek damages for its costs incurred (such as extending the validity of its performance bond) in connection with the government's postponement of construction because the contract contained a "no damages for delay" provision. In its irritation in hearing the contractor's claim, the court reprimanded the contractor by pontificating that "men who take million-dollar contracts for government buildings are neither unsophisticated nor careless" and thus the court presumed that the contractor had protected itself against the cost of delays caused by the government when it "quoted" the contract price to the government in the first place.

2   See *In re Regional Building Systems, Inc.*, 320 F.3d 482 (4th Cir. 2003), holding that a general contractor was not responsible to a subcontractor for the owner's delays when there was a "no damages for delay" provision contained in the subcontract.

3   See *Gheradi v. Trenton Board of Education*, 53 N.J. Super. 349 (App. Div. 1938), ruling that delays contemplated at the time a contract is signed do not amount to "active interference."

4   See *Pellerin Construction, Inc. v. WITCO Corporation*, 169 F. Supp. 2d 568 (E.D. La. 2001), in which a subcontractor was not able to invalidate releases of liability it had given to its counterparty during the project's construction and then bring an "active interference" claim against its counterparty when construction was complete.

Chapter 15

Unexpected Market Conditions

Forecasts are not always accurate. A change in macro-economic, political or social conditions can affect a project's viability. The owner must do its best to protect itself against uncertainties by reserving the right to discontinue, suspend or slow prosecution of the work. For instance, in the case of power or LNG projects, a serious slump or a sharp spike in gas prices could make building a "peaking" power plant or LNG terminal much less viable. Many technologies have their "day in the sun." For some, the shade may be only temporary cloud cover; for others, it may mean sunset. In 1620, King James commissioned his advisors to report on fuel supply and they concluded "no wood, no kingdom." Thus, the British Empire turned to coal to replace wood. Later, it turned to oil to replace coal and British Petroleum (perhaps the most successful economic venture ever) led the search for oil. Natural gas has finally replaced oil as the preferred fuel. New techniques to unlock trapped natural gas may propagate natural gas's reign. Nuclear, solar, wind, wave and tidal energy remain viable but expensive options for electricity. Yet, although it is like building a dinosaur, coal power plants are still built because the particular situation may require it. Sometimes technological change takes decades, sometimes public opinion can change the future for a technology in a few hours. If a technology change (whether from global public perception or loss of a lawsuit regarding a permit) occurs after notice to proceed has been given to the EPC contractor, the owner may want to scale down its investment (for example, in the case of a power plant by reducing the size of its plant—perhaps by installing less generation capacity than originally anticipated).

What if a country's political situation or its currency become unstable? The owner may want to curtail spending on the project by slowing work in order to wait and see if the situation is likely to stabilize. Worse yet, what if (as has basically been true from California to Turkey) the government or utility offtaker simply announces that it will, or will attempt to, unilaterally terminate the concession or offtake agreement for the project?

In all of the above cases, the owner needs the flexibility to retard or stop its spending while it is addressing and assessing implications of external threats to its project's profitability.

The Owner's Right to Suspend or Terminate Work

The EPC contract should contain a mechanism for handling macro-economic and geopolitical complications. It should give the owner the right to suspend or slow the work by giving the EPC contractor a written notice, whereupon the EPC contractor must take adequate precautions to preserve and protect the portion of the work suspended or slowed and also make a reasonably detailed account of the suspended work. All this will be done at the owner's expense. Once the affected work (or the original pace) recommences, the EPC contractor will be entitled to an equitable adjustment for increased costs associated with the owner-imposed delay.

If the predicament is dire, the owner may need to terminate all or part of the work permanently. The EPC contract should provide that if all the work is terminated, the EPC contractor will be paid for the cost (not value because that may be difficult to reach agreement upon with the EPC contractor) of any work performed for which payment has not yet been received plus any reasonable cancellation and de-mobilization costs (including costs of the EPC contractor's premature termination of any of its subcontractors) incurred by the EPC contractor as a result of the owner's termination.

Whether or not the EPC contractor should be entitled to its expected profit in connection with termination of some or all of the work is a subject for debate. The first question is whether or not the EPC contractor should be entitled to all, or just a portion, of its expected profit on the EPC contract. If the owner has terminated the project at an early stage, should the EPC contractor be entitled only to some portion of its expected profit depending upon when in the project lifecycle the project has been terminated by the owner? There are even more basic questions. What was the amount of profit that the EPC contractor expected at the time it signed the EPC contract? What would have been the amount of its profit at the time the EPC contract was actually terminated? Both questions are difficult for the owner to answer without the EPC contractor's cooperation and full disclosure. Expected profits will depend upon what the EPC contractor's original expectations were about the project and whether or not they were realistic. Profit level will also depend upon whether or not the EPC contractor was able to stay within its budget during the project's progress. Unfortunately, all of these are items the owner will have great difficulty surmising accurately. The owner will have to make its own estimates and corroborate them with the EPC contractor's statements.

To avoid a long and uninformed discussion at the negotiating table about termination, the most typical compromise for the parties is to agree upon a "cost of termination" schedule that specifies the amount to be paid by the owner to the EPC contractor in the case of termination depending upon the time that has elapsed since notice to proceed was given to the EPC contractor by the owner.

Partial termination of the work (relatively more likely than a full termination) is much more complicated because its effects on the EPC contractor's work and costs cannot be evaluated at the time the EPC contract is executed (because at that time it is not known what portion of the work will be terminated). For this reason, neither a predetermined charge nor a formula seems appropriate. It is usually best to handle these partial termination cases as equitable adjustments under the equitable adjustment provisions of the EPC contract so that the contract price can be adjusted if a partial termination occurs.

In all of the above cases, the EPC contract should make clear that the EPC contractor's sole remedy for suspension or termination of the work by the owner is expressly stated in the EPC contract (either as a formula or as equitable adjustment) and the EPC contractor should not be entitled to lodge a claim against the owner on account of a termination whether for lost profits or consequential damages or anything else. The owner does not want to go through the process of calculating and making termination payments only to be summoned to court by the EPC contractor's complaint that the parties never intended that this payment would compensate the EPC contractor for its lost profits and that the owner should therefore also account to the EPC contractor for the EPC contractor's lost profits in addition to its termination costs.

Chapter 16

Testing and Completion of the Work

As construction and installation activities draw to a close and the silhouette of the facility begins to emerge, a battalion of startup and commissioning experts will replace the legions of erectors and installers at the project site. Once all the equipment is bolted into place and interconnected to the facility's controls, someone must determine when it is safe to see whether the facility "starts." This chapter discusses the key contractual concepts that are involved once a facility has been erected (see Figure 16.1).

Many components are tested individually before they leave their factory (such as transformers and pumps). However, many large items cannot be fully tested before they are shipped. For example, in the case of a power plant, combustion gas turbines sometimes are "spun" but not tested with their generators connected because the shops in which they are fabricated usually do not have any place to discharge the huge electrical charge that they generate. Obviously, whether or not items are tested individually, they must be tested as an integrated system.

Mechanical Completion

Customarily, once all equipment has been installed and is ready for commissioning, the EPC contractor is said to have achieved "mechanical completion" because the facility is mechanically assembled and, in theory, ready to operate once equipment is tuned and calibrated in accordance with manufacturers' recommendations and the demands of ambient conditions at the site. In the case of a power plant, altitude, temperature, humidity and barometric pressure all affect the plant's output.

Startup and Commissioning

Startup and commissioning can take weeks or months. The EPC contractor must allow adequate time for all this testing when it is developing the schedule for the project (see Chapter 11). For example, a relatively simple boiler tube problem could take 10 minutes to identify but weeks to correct because refabrication of the inadequate tubing could take a month or more depending upon shop availability and shipping schedules. Worse yet, a system that is not functioning properly may have to be redesigned and reinstalled.

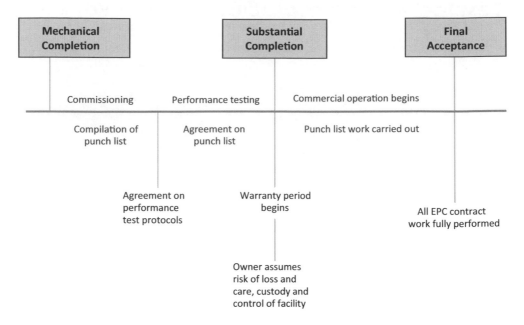

Figure 16.1 Completion timetable.

The Punch List

During the commissioning program, the EPC contractor and the owner will tabulate what work is not completed but should have been completed by the time of commissioning. This tabulation is commonly referred to as the "punch" list. It is a list of work that remains to be performed or must be redone because it was not performed correctly. Generally, work identified on the punch list will not jeopardize operation of the facility but nonetheless must be effectuated before the project can be considered complete. Items on the punch list run the gamut from defective welds to non-delivery of the *in situ* (on site) crane to be installed for future lifting of heavy equipment during maintenance overhauls.

The EPC contractor and the owner often have little problem agreeing on the punch list items. However, the owner and the EPC contractor often disagree on the value or cost of items on the list. Sometimes, negotiation of the punch list can be an ordeal for an EPC contractor. Negotiation of the punch list can be used as an opportunity for a disorganized, inexperienced or unscrupulous owner to extract economic concessions from the EPC contractor because the EPC contractor may still be awaiting payment of a substantial sum under the EPC contract. As the owner's operating personnel begin to survey the facility before it is handed over to them, they may have many requests, either as afterthoughts or because they had misconceptions of what items actually were included in the EPC contractor's scope and now those items are, in their judgment, missing. Arguments can develop, for instance, over whether or not a valve should have a staircase and an access platform for easy access by operating personnel or whether the operating personnel are instead expected to use a man-lift to access the valve. Other issues can be questions of landscaping, site restoration, access roads or paving of fire lanes. Of course, all these debates will arise only if the functional specification is either ambiguous or lacks detail. Again, the issue of the intent of the parties may have to be used to resolve these disputes. If the parties' intent was for the EPC contractor to design and build the facility in accordance with prudent practices, the question is: do prudent practices, for example, require that an access platform is necessary for access to a particular valve, or that a man-lift is an acceptable and prudent means of accessing the valve given added safety concerns over the instability of the man-lift versus a staircase and the expense of purchasing and maintaining the man-lift? For a court or an arbitrator, all of the foregoing are questions of fact on which expert testimony is likely to be sought.

Aside from preparing a precise functional specification, the EPC contractor's best strategy for neutralizing the holdup value that owners can exert in refusing to agree upon the punch list in a prompt manner is to include a provision in the EPC contract that states that the EPC contractor's agreement to placing an item on the punch list does not preclude the EPC contractor from making an equitable adjustment claim that such work was never within the scope of the functional specification. Therefore, the EPC contractor should be entitled to an equitable adjustment because the "punch list" item may really just be extra work that is properly characterized as a directed change being given by the owner and not incomplete performance of the EPC contract by the EPC contractor. If a provision like this is included for the EPC contractor's benefit, the owner, to protect itself from unexpected claims, should make certain that the EPC contract provides that the EPC contractor can make only an equitable adjustment claim for an item if the EPC contractor expressly makes a written objection to the owner about the item at the time the punch list is agreed upon by the owner and the EPC contractor.

The owner should be cognizant of the EPC contractor's interests during the commissioning phase. After a long engineering and construction cycle, the EPC contractor is likely to be suffering from project fatigue and eager to get its final payment, which may contain most, if not all, of its profit (depending upon whether the EPC contractor was able to stay within its budget for the project) and move on to other projects. An EPC contractor may do only just enough testing and tuning to make sure that the facility passes the owner's performance tests.

Output during Commissioning

For the most part, while an EPC contractor will generally not dawdle in concluding commissioning, packing up its equipment and vacating the site, the owner should protect itself against "hidden" costs that can arise if the EPC contractor runs into difficulties during the testing of the facility. For instance, in the case of a power plant, the owner usually will not be paid the full tariff for electricity the power plant produces during commissioning because this "infirm test power" is often not stable and the power station cannot be relied upon to remain in continuous operation, so the local electric utility may need to keep another power station on "hot" standby. Therefore, the utility cannot include this power in its daily scheduling of power generation resources because the plant may fail to start up or even shut off unexpectedly (usually referred to as a "trip") once it has been permitted to generate electricity by the utility dispatcher. Sometimes the owner will not be paid at all by the utility for power generated during testing of the plant. With all this in mind, the owner must insure that its own fuel costs will be covered during the power plant's testing period, especially since the power plant may not yet be capable of operating at the efficiency (heat rate) that the EPC contractor has promised to achieve. The owner can protect itself in several ways, either by requiring the EPC contractor to pay the owner for any of the owner's fuel or feedstock costs or by agreeing that the owner will be required to spend only a certain limited amount on fuel and feedstocks and, once this limit has been reached, the EPC contractor will pay (or reimburse) the owner for any costs above these levels.

Performance Testing

Once commissioning is complete, the facility is ready for the performance tests that will be conducted in order to evaluate whether or not the facility meets the owner's specifications. Ideally, the minimum performance thresholds of the facility and the performance tests' reference conditions, corrections and duration should all have been agreed upon in advance and included as an exhibit to the EPC contract. Often, however, the test protocols cannot be agreed until after the EPC contractor has completed some of the design of the facility and selected the equipment that it will install in the facility. While the practice of agreeing upon test protocols only after the EPC contract has been signed is frequently unavoidable, the owner should keep several things in mind when it does eventually agree upon protocols. Testing on an individual train and also a combined facility basis is always preferable for the owner if the facility contains more than one train. For instance, if a power plant has two combustion turbine units but is tested as a complete power plant only when both combustion turbines are operating, an overperforming combustion turbine unit could mask a problem with the other, underperforming combustion turbine unit. What will happen when the overperforming combustion turbine unit undergoes maintenance and the owner is left running just the inferior combustion turbine unit? In such a case, the owner might be assessed penalties by its offtaker. Furthermore, a facility should not just be tested by trains individually and then test results summed because this summation could mask a problem with a facility's common facilities. Again, in the case of a power plant, a shortfall in cooling capacity or excess "parasitic" electricity load demands (such as electricity to drive pumps) may be apparent only when all units are running at maximum capacity.

Unit Testing

A facility's output at reduced levels should always be tested on both a train-by-train and a full-plant basis to determine that pollution emission and operational efficiency levels can be met at partial loads because the facility may actually be forced to operate in this mode for economic or other reasons. Especially in the case of a power plant, for system reliability reasons, the power grid operator will always prefer to have two combustion turbine units of a two combustion turbine power plant running at half load rather than one unit of the power plant running at full throttle. This is true because a unit that is in operation can respond more quickly and reliably to an increase in the grid's electrical demand than a unit that is cold and must be started, which could take minutes, hours or days. In fact, most power plant "trips" usually occur during startup. Sometimes owners will even choose to run units at low loads during the night (instead of turning them off) so they do not have to risk not being able to start up the combustion turbines for the power grid's surge in electricity demand in the morning. However, from an efficiency point of view, owners will usually prefer to run one unit at full load rather than two units at half load because half load is typically inefficient from a fuel usage point of view (akin to the concept of higher miles per gallon for highway driving vs. lower miles per gallon for city driving). Another important subject for the owner and EPC contractor to agree upon is how many output levels will be tested and how many trips of a particular unit and the facility will be permitted before a test is "failed" (for example, no more than one trip in any 48-hour period).

Power Plant Degradation

Another issue that can lead to disputes in the case of power plant testing is equipment degradation. Output of combustion turbines (but generally not steam turbines) degrades noticeably during their first few hundred hours of use. For example, if the EPC contractor conducts 95 hours of commissioning tests, should the degradation that occurs during this period be taken into account when measuring the plant's performance against the performance promised by the EPC contractor in the EPC contract? Combustion turbine manufacturers usually have degradation charts for each of their turbine models. These degradation charts (usually plotted as a curve) are usually based on the manufacturer's experience with the operational history of its combustion turbines. While the degradation of any particular combustion turbine unit may vary from the curve supplied by the vendor, these empirically derived curves are useful for predicting the level of degradation that can be expected for a particular turbine. For example, a manufacturer's degradation curve may predict that a particular gas turbine model's electrical capacity will degrade by about 1 percent after 100 equivalent operating hours of use.[1] Consequently, if a manufacturer has promised that its gas turbine will produce 100 MW but the gas turbine has undergone 100 hours of testing, the manufacturer's guarantee will usually be adjusted to 99 MW to account for this expected degradation. If, however, the gas turbine has degraded by 3 MW during the 100 hours of testing, the manufacturer will usually be responsible for the two additional MW of degradation and either be required to pay liquidated damages to the owner or somehow reverse the degradation (which could be much more costly than paying liquidated damages).

It is important to establish and make clear in the EPC contract whether the numerical values of performance guarantees refer to the combustion turbine in its "pristine" state at its first firing or the combustion turbine after it has been fully commissioned and tuned. The degradation of the combustion turbine during its commissioning should probably not be the owner's problem. It should be an issue to be dealt with between the EPC contractor and the combustion turbine's manufacturer.

Performance Curves

Finally, degradation curves should not be confused with performance curves. Degradation curves predict how a combustion turbine will perform over time as it wears and becomes fouled by fuel residue. Performance curves (or power curves in the case of a wind turbine), on the other hand, predict how a combustion turbine (or wind turbine) will perform under different ambient site conditions given the power plant's location (for example, at sea level). Performance curves usually have to be prepared for each individual combustion turbine (as opposed to a model of a combustion turbine) based upon where the plant is located and what accessories have been added to the combustion turbine unit (such as dry, low nitrous oxide [DLN] combustors). Although it is typical for the EPC contract to provide that the performance curves will be developed during the course of the EPC contractor's work (because their development requires information usually not available at the time the EPC contract is executed), it is in the owner's best interest to have these curves developed as soon as practicable to avoid misconceptions about plant performance and thereby enable the owner to request appropriate design adjustments, or enhancements, if necessary, so the power plant can meet the owner's expectations. Unfortunately, such changes can necessitate a costly change order.

Test Measurement Locations

Another source of testing disputes can be the measurement locations. A power plant owner should insist that electrical output be measured at the high-voltage side of the "step-up" transformer after the voltage has been increased by that transformer. This increase in voltage level allows electricity to travel longer distances with lower line losses. This measuring point will also automatically take into account all the power plant's parasitic load demands and transformer losses, which is why the power plant's revenue meters for electricity sales to the offtaker are usually located at this point or beyond it. On the other hand, the EPC contractor will usually prefer to measure output at the plant's generator terminals and derive parasite loads and transformer losses separately.

Extrapolation

An owner should also understand whether a measurement at a designated measurement point will actually be taken or simply extrapolated from a calculation of measurements taken at other locations. For instance, combustion turbine exhaust or mass flow is usually an important measurement point in combined cycle power plants because exhaust energy determines how much steam the heat recovery steam generator will be able to produce for injection into the plant's steam turbine. However, generally, there is no meter sitting in the 950°F gas turbine exhaust to measure this. Rather, a calculation is made from other measurements made in the plant's systems, such as its steam turbine output. If "derived" measurements will be made, the owner should understand and agree upon the methods for calculations and derivations. Obviously, there are many more instances in which measurement points and calculations can affect results. To avoid disappointment and dispute, non-technical professionals involved in the negotiation of an EPC contract should do their best to understand the principles contained in the EPC contract's testing protocols. If they cannot understand the underlying test principles, the protocols are probably not clearly written and should be rewritten. After all, in a worst-case scenario, it will be an arbitrator, judge or jury trying to decipher and understand the test protocols based upon what technical experts wrote in shorthand because they thought that all were in agreement.

Offtaker or Facility User Requirements

The owner should also try to incorporate the requirements of its offtaker or facility user into the facility's performance test's reference conditions and protocols. For instance, in the case of a power plant it does the owner little good to agree in the EPC contract that the power plant will be tested at a 0.9 power factor when the power purchaser will insist on testing the plant at a 0.85 power factor because 0.85 is the power grid's operating requirement and equilibrium state.[2]

Correction Curves

The owner must also recognize limitations inherent in correction curves, especially temperature correction curves. In the case of a power plant designed for peaking service in the summer season, the plant will be in service primarily when the temperature is relatively warm. Thus, the plant should be tested in hot weather, not cold weather "corrected" by extrapolation to hot weather conditions. If the EPC contract provides that a power plant must be ready for service in August but delays push the in-service date into December, it is not wise for the owner to permit performance tests to be conducted in December instead of August. Many systems' performance may be enhanced by relatively cold weather, such as the cooling tower, which generally is more effective at dissipating heat from the plant's hot water when the temperature is relatively cold but may not function as required when the temperature is relatively warm and may not have been designed with enough surface area to remove heat from the water effectively in the summer. For power plants, correction curves for combustion turbines may not be accurate over very broad temperature ranges. In circumstances where a power plant is not completed during the season that it is expected to perform at its peak performance, the owner and the EPC contractor should agree in the EPC contract that preliminary testing will be carried out on a provisional basis until results can be verified at a later date when atmospheric conditions are similar to those in which the power plant is designed to operate.

Deemed Acceptance

"Deemed acceptance" or "deemed completion" of a facility may be necessary in many circumstances. In fact, experienced EPC contractors will usually insist that the EPC contract contain a provision dealing with circumstances that may arise to prevent or postpone performance testing. For example, in the case of a power plant being constructed near a natural gas production field to supply power to an air separation unit that will produce nitrogen for injection into the natural gas field (so the gas field can be exploited more effectively), suppose the EPC contractor completes the power plant but the natural gas to fuel the power plant is not yet available so that the power plant cannot be tested. In this case, since the EPC contractor is (in its opinion) finished with its work, the EPC contractor would like to have the power plant "deemed" complete so that it can be paid, vacate the site and commence the warranty period. Acquiescing that a power plant is complete before it has been tested is not a good idea for the owner. To address this type of circumstance, a compromise can usually be reached in the EPC contract that will provide that the EPC contractor will continue to maintain the facility in "cold" storage at the owner's expense for some period. If performance testing cannot commence within a certain period, the facility will, in fact, be deemed complete by the owner (because the EPC contractor will not agree to postpone testing of a facility indefinitely). In the instance that the EPC contractor is delayed in carrying out performance testing, the EPC contractor usually will insist that it receive some portion (or all) of the payments it would have received once performance testing had been successfully completed.

Of course, the owner is ill-advised to make the final installment payment for a facility that has not been tested, and several arrangements can be made to deal with this situation. These arrangements include escrows, letters of credit and partial payments. It is important to consult with the project's insurers in drafting "provisional acceptance" provisions because insurance policies must accurately reflect the stipulated responsibilities of the parties while a facility is in "suspended animation" or being operated until testing is possible because builder's all risk insurance generally terminates once a facility is placed into commercial operation. That is true whether or not performance testing has been completed. Thus, modifications to insurance policies may have to be made to continue coverage under such circumstances.

Many events can delay performance testing. Consider a mine-mouth power plant. Instead of transporting coal to a power plant, it may be more economical to build a power plant near its fuel supply and build electric transmission lines to connect the power plant to the "load pocket" where the electricity is needed by customers. In designing the power plant, the sponsor may lay out the equipment so that additional generating units can be added in the future if electricity demand grows. Obviously, it will not spend the money to build generation units if they are not needed but in building the transmission line the owner may decide to "oversize" the cable so that the cable can handle the increased electrical power of the additional generating units that may be added in the future. Consequently, no modifications to the transmission cable will be required if the additional generation units are ever built. The owner may have made this decision because a cable with a higher megavolt ampere rating (which can transmit more power) may not be much more expensive than installing a cable with a lower rating, and adding a second cable in the future might be very expensive at that time. Thus, the owner might decide to put some capital at risk and install a larger cable in the hope that this additional capital cost gamble will pay off in the future. Unfortunately, there is no way for the "oversized" cable to be tested until sometime in the future, if and when additional power generation units are built. Since it is unlikely that the EPC contractor will agree to undertake a performance test many years in the future, the owner and the EPC contractor are usually able to agree upon some sort of factory or field testing with an extrapolation that can serve as a proxy for currently unattainable conditions. Another example of such an over-capacity approach would be pipelines built with more capacity than needed but designed so that the owner can add compressors or pumps in the future to avail itself of additional capacity if it is ever needed.

Reliability Testing

Performance tests can generally be divided into two sub-sets. The one tests reliability of the facility and the other tests performance. In power plants, reliability tests often range from 48 hours to about 172 hours depending upon the type of power plant being tested. During this period the power plant's ability to run continually while responding to different load conditions will be evaluated. Performance testing is usually done during a 2–4-hour period in which electrical output and efficiency are tested at different throttle settings. Performance testing is often done coincidentally during an interval of the reliability test once the power plant has been brought up to "steady state" conditions so that its equipment is operating at optimum design levels. Then readings will usually be taken at the beginning and end of specified intervals (usually every 15 minutes) and averaged in an attempt to minimize temporary fluctuations.[3]

Ambient Air Temperature

Many facilities (especially power plants) that operate in warm climates will be equipped with "inlet" chilling systems that will cool (and sometimes "foggers" that will dampen) air before the air is used in their process. At a power plant, air that has been cooled and dampened before it goes into the combustor will increase electrical output because more energy can be released in the combustion process. Thus, inlet air temperature can mean different things (that is, ambient air temperature at the site or the air temperature after air has passed through the air chiller) and it is best to be clear at what locations temperature adjustments will be made. It should also be specified that no adjustment to test results will be made if the air temperature is cooler than that of the agreed-upon air temperature reference point. Thus, if a power plant is supposed to be tested at a minimum of 86°F and the temperature at the time of the test is 88°F, no adjustment to the electrical output should be made for the EPC contractor's benefit because the plant should meet the required electrical (and thermal) output levels at any temperature above 86°F.

Performance tests will be performed while the owner's operating staff are operating the facility under direction of the EPC contractor. Before the tests are conducted, the owner's personnel will have had to complete the EPC contractor's training program. Sometimes concerned EPC contractors impose a specific obligation on the owner in the EPC contract to supply an adequate number of appropriately skilled operators to carry out the performance tests.

Once performance tests have been concluded, it often takes days or even weeks to gather, confirm and correct all the test data. Evaluation of test results can include items such as fuel, product and waste assays (which usually must be sent to off-site laboratories for analysis). The EPC contract will give the owner time to assimilate test results and conditions (which, if the owner has financed its project, usually must be communicated to their engineer so the lenders can intervene if they believe that they have identified a problem the owner has not) before the owner agrees that the performance tests have been successfully completed. If the owner does not "sign off" on the test results, the owner must explain the rationale for its failure to agree that the facility has passed the performance tests. The tests will be repeated once the EPC contractor believes that the problem has been rectified.

Performance Tests for Offtakers and Lenders

The owner must be cognizant that there may be three separate batteries of testing necessary for the owner to determine that all its contractual obligations concerning the project have been satisfied. These three separate regimes may or may not be capable of being conducted simultaneously. For strategic reasons it may be wise to separate them entirely.

The first series of tests will be those the owner requires of the EPC contractor under the EPC contract to ensure that the EPC contractor has built the facility that the owner expected. This will be the most thorough set of tests. Typically, these tests will encompass testing of individual pieces of equipment and systems (because performance of facility components will affect the owner's operating costs) as well as the facility as a whole.

If the owner has financed its project, the second battery of tests will be those necessary to satisfy the owner's lenders that the facility (which is the collateral for their loans to the owner) is ready to perform as required for the owner to run its business profitably. Generally, these tests will be the same as those that the owner imposed upon the EPC contractor and typically the lenders' engineer will merely witness and examine the results of those tests. However, in certain circumstances, such as when lenders believe that the owner has done a poor job in negotiating the EPC contract or when the owner and the EPC contractor are affiliates (because the EPC contractor indirectly owns some portion of the owner's project), the lenders may require that the facility satisfies additional tests.

Finally, the owner's offtaker or user is likely to have its own testing requirements for the facility. Tests imposed by the offtaker or user will vary depending upon the needs and sophistication of the offtaker or user. For instance, suppose a power plant is linked to a large electricity grid and the offtaker is a very large energy trading company with many electricity purchase contracts in many markets. This offtaker may have few, if any, requirements other than that the owner deliver electricity to the delivery point required by the offtaker. To the extent the owner fails to make deliveries as requested, the offtaker may be content to collect predetermined liquidated damages for shortfalls in supply or to recover the costs that the offtaker incurs for purchasing replacement power to replace electricity the owner failed to deliver. In juxtaposition, however, suppose the owner's customer is an electricity utility located on an island that has only six power plants and therefore the utility is relying upon the owner's power plant as a resource to support (and not disrupt!) the island's power grid whose users may literally be in the dark because no other power station exists to supply replacement power. In this case, the owner's customer is much more likely to require rigorous reliability testing to try to determine that the owner's power plant can handle grid disturbances (such as dropping voltage, etc.).

In all of the above cases, the owner may not want its offtaker to witness (or analyze the results of) the testing done by the EPC contractor (which may expose inadequacies in the power plant). Therefore, it may be better to have the EPC contractor conduct all its performance tests before the owner begins to conduct performance tests for its offtaker, rather than perform both sets of tests simultaneously. Otherwise, in connection with the offtaker's monitoring of the extensive testing that the owner requires of the EPC contractor, the offtaker might observe something it believes could compromise the reliability of its own operations and therefore the owner may suddenly find that its customer will ask or require the owner to expend additional funds to make changes.

In situations in which the owner's customer is truly depending upon the power plant's physical output for the stability of the customer's grid system or production process, the owner should take this into account in the power plant's design criteria. For instance, suppose the grounding grid for a power plant is designed with the resistivity and number of earth spikes recommended by the relevant industry electrical code. Suppose, further, that one day the owner's plant is struck by lightning and the plant's electronic controls are "fried" by the electrical surge caused by a lightning bolt and require two weeks to be replaced, during which the power plant remains out of commission. Then, suppose the owner sends its offtaker a notice of *force majeure* requesting that the owner be excused from the owner's power delivery obligations for the two weeks that the plant was out of commission as a result of the lightning strike. Suppose further, in response to the owner's notice, that the owner's customer, instead of acknowledging the *force majeure* claim made by the owner, denies the *force majeure* claim asserted by the owner on the basis that the owner did not adhere to prudent industry practices in designing a lightning protection system for a power plant located on an island because the owner should have utilized a dissipation scheme with a lower resistivity and more earth spikes in the power plant's grounding grid, and therefore, the utility sends a bill to the owner assessing liquidated damages against the owner claiming that the power plant experienced an "unexcused" outage for two weeks and the utility posits that the mere fact that a lightning strike caused an outage of the power plant is *prima facie* (clear) evidence that the lightning protection system of the power plant was inadequately designed. Unfortunately for the owner, its offtaker has a credible case in these circumstances. Therefore, it is important for the owner to be able to demonstrate in cases like this that the owner did, in fact, give due consideration to the location and strategic importance to the offtaker of the power plant. In fact, whether or not the owner has a warranty claim in this case against the EPC contractor may depend upon the intent clause in the EPC contract if the EPC contract failed to specify with particularity the requirements of the lightning protection system. This is why (as was discussed in Chapter 6) the intent clause of the EPC contract should specify the location and type of service that a facility will be expected to provide.

"Inside-the-Fence" Offtakers

In cases of a so-called "inside the fence" power or steam customer such as a refinery or steel mill, the customer may impose different performance test requirements upon the owner than would a utility customer. For instance, if an owner's facility is providing power and steam and/or chilled and hot water to a customer such as a university or hospital, the customer may want to make sure that the facility can meet the customer's so-called "coincident peaks" in all seasons and at all times of the day when the demand for all the facility's outputs (such as electricity, steam, hot water and chilled water) are high and the owner cannot compensate for one poorly operating system at the expense of another facility system. For example, one way to control nitrous oxide emissions from a power plant is to inject steam into its combustor. The steam has the added benefit of increasing the efficiency of the combustion, thereby increasing the electrical output of the combustion turbine. So, on occasion, when higher power output is necessary, additional steam can be channeled into the combustor. However, if more steam is channeled into the combustor in order to increase electrical output, less steam is available to meet the customer's steam and hot water needs. Thus, the customer may require that steam injection into the combustor during performance testing be limited to that amount which is required to attain the nitrous oxide emissions level contained in the power plant's air permit. Sometimes these operational "balancing acts" (such as the diversion of steam) may not be an issue when the owner's plant is placed in service but can become critical in years to come if the customer has future expansion plans for its campus or production facility.

Merchant Power Facilities

Finally, in the case of a so-called "merchant" power plant in which the owner does not have a utility customer but will be selling its electrical output on the open market, its power plant will usually have to pass the operational tests imposed by the local system operator of the power grid to participate in the local power pool. These tests are customarily the least burdensome from the standpoint of intricacy and difficulty to satisfy. Many power plants typically run on what is known as automatic generation control (AGC), whereby the grid system operator (usually referred to as the "dispatcher" or "scheduler") actually manipulates the output of the power plant remotely from the dispatcher's command center and the power plant is taken off AGC only when it is in startup or shutdown mode or in an emergency.

Many power pools provide "bonuses" for power plants that can offer so-called "ancillary services" such as "quick load pickup" or "spinning reserve" in which a power plant already synchronized with the power grid (that is, currently on-line) can increase electrical output rapidly in response to increased demand from users on the grid and thereby prevent grid frequency from dropping and potentially damaging generators of other connected power plants because their units must work harder to meet the sudden electrical demand. In fact, generation stations often set circuit breakers to automatically open and disconnect a power station from the grid if the grid's frequency drops below a certain level in order to protect their generators from internal damage as they strain to try to suddenly meet increased demand.

Other such ancillary services may include provision of voltage area regulation (commonly referred to as "VARs"), which is reactive power to help stabilize grid conditions.[4]

Another ancillary service is "black start" capability, which allows a power plant to start without backfeed power from the grid because the power plant has its own batteries or diesel generators on site so that the plant can start up in emergencies.

In conclusion, the owner should make certain that the successful completion of all tests that the owner's facility will undergo for any of the owner's stakeholders (such as permitting agencies, lenders and offtakers) are the EPC contractor's responsibility. Generally, an EPC contractor will resist this responsibility on the basis that it cannot control the decisions of third parties as to whether or not the facility has satisfied a third party's testing requirements. This may seem true. It does not seem equitable for the owner to withhold the EPC contractor's final payment until the owner's customers (and lenders if the owner has financed its project) are satisfied with the facility's performance, but the truth is that the facility is absolutely worthless to the owner unless the offtaker or user agrees that the facility is ready for commercial service and will commence payments to the owner under the offtake or use agreement. Of course, it is unfair for the owner to "hold up" the EPC contractor if the offtaker or user is arbitrarily withholding confirmation that the facility is ready for service or if the offtaker or user is denying that the facility is ready for commercial operation because the owner has failed to satisfy a condition of the facility being placed into commercial service that is unrelated to the EPC contractor's performance (such as the owner's failure to procure property insurance in conformance with terms of the offtake or use agreement).

Often, the most expedient compromise with the EPC contractor regarding requirements of third parties is to require that the facility must satisfy all the objective and technical requisites contained in the testing regimes of these third parties but not the prerequisites that are within the discretion of the third parties (such as the delivery of certificates of approval of work). For example, in the case of a power plant, an example of an objective requirement might be that the facility must be capable of continuing to operate during a three-phase fault on the power grid lasting less than one second. An example of a "discretionary" requisite imposed by an offtaker might be the written acknowledgment by the offtaker that the facility is safe to operate. If the owner has financed its project, lenders may actually require the satisfaction of all the offtaker's conditions (objective or subjective) because the owner will have no income to support the lenders' loans until the offtaker or user commences payment to the owner under the offtake or use agreement because the offtaker is satisfied that all conditions for payment under the offtake or use agreement have been satisfied. Thus, even a relatively small item that leads the offtaker or user to assert that the facility cannot be placed into commercial operation (such as a faulty data phone connection that prevents real-time access to the facility's operation but certainly is not needed for plant operation) has the potential to jeopardize the owner's economics if the project's revenue stream does not commence when the owner (and if it has financed its project, its lenders) anticipated. The problem could be still more serious if the offtaker or user and owner become involved in a dispute over a "large-ticket" item such as whether or not the interconnection facilities have been designed properly. At the same time, if the owner has financed its project, the owner must be wary lest the lenders claim a default under their loan agreement with the owner because they may not believe the dispute will be resolved in the owner's favor.

For instance, take the case in which an owner has built an overhead transmission line through a desert to connect its power plant to a transmission grid or the case in which an owner has laid a transmission cable under a river to connect to a transmission grid. Suppose, in the first case, that during the testing of the power plant it turns out that dust storms in the desert cause resistors on the transmission line to become covered in sand and then they often explode as a result of their heat, which can no longer be dissipated into the air because they are encaked in sand, and therefore the power purchaser concludes that the transmission line will be unreliable and that the owner should have buried the transmission line instead of stringing it on poles. Or, in the case of the underwater cable, suppose the cable has been buried too deeply in the river bed by the EPC contractor and as a result the heat created by the electrical current in the cable cannot dissipate as anticipated because the cable is lying on bedrock that does not absorb heat quickly enough and therefore the cable is not capable of transmitting the electrical load that the owner anticipated. Unfortunately, in either case, the power purchaser may claim that the owner did not design its power system in accordance with prudent practices and therefore the power purchaser may not be willing to acknowledge that the owner's facility is ready for commercial service or may refuse to pay the full tariff that the owner has negotiated in the offtake agreement until the owner corrects the problem (that is, buries the cable underground in the first case or decreases the cable's burial depth in the second case). In these cases, if the owner has financed its project, its lenders may not have any interest in waiting to see what will happen and invoke the default provision under their loan agreement with the owner before the dispute is settled. In fact, even if such a dispute is settled without dire financial consequences to the owner, the owner may need to apply for new burial permits or subsurface rights of way, which could require new environmental studies, take years and perhaps be unsuccessful. The risk of encountering these types of unanticipated problems (in addition to simple inadequate design and construction) is why many types of lending institutions (such as pension funds and insurance companies) generally avoid lending to projects before construction is complete and the project has been placed into service. Conversely, this is why lenders who are comfortable assessing these types of construction risks can charge a premium for making construction loans to projects.

Substantial Completion or Provisional Acceptance

Once performance tests have been passed and the facility performs in accordance with the EPC contract's specifications, the owner will check to see whether the EPC contractor has achieved "substantial completion" or "provisional acceptance" under the terms of the EPC contract. "Substantial completion" or "provisional acceptance" are construction industry terms of art and are often misused and misunderstood. "Substantial completion" or "provisional acceptance" does not refer to the physical completion of a facility and satisfaction of its performance tests. It refers to substantial completion of the EPC contractor's responsibilities under the EPC contract, which go beyond completing a facility that can meet output and efficiency guarantees. Substantial completion or provisional acceptance involves completion of many non-construction-related activities such as delivering as-built drawings, operating manuals and releases of liability from subcontractors. In the case of a power plant, substantial completion or provisional acceptance may also involve passing tests imposed by the local grid system operator and delivery of electric relay and protection schematic diagrams to the grid operator. By the date of substantial completion or provisional acceptance, the EPC contractor should also have paid the owner any liquidated damages it owes the owner for delay in completion of the facility or for the facility's retarded performance.

Substantial completion or provisional acceptance means that a project is ready to enter its operational phase. Generally, if the owner has financed its project, its lenders will not extend the maturity of their construction period loans into long-term loans unless the EPC contractor has met all the requirements of substantial completion or provisional acceptance contained in the EPC contract, not just requirements related to physical operation. Thus, the EPC contract should contain a very detailed list of the requirements that the EPC contractor must satisfy in order to achieve substantial completion or provisional acceptance.

Substantial completion or provisional acceptance will typically entitle the EPC contractor to a large (and often final) milestone payment and commence the warranty period.

The owner should also include a provision in the EPC contract that will bar the EPC contractor from filing claims for equitable adjustments after the date of substantial completion or provisional acceptance. This is prudent because the owner can be at risk if the owner makes a large milestone payment to the EPC contractor and therefore the owner believes that the final project capital cost has been fixed only to learn later that the EPC contractor is filing an equitable adjustment claim.[5] The owner could even attempt to require that all outstanding equitable adjustment claims be satisfactorily resolved as a precursor to achievement of substantial completion or provisional acceptance by the EPC contractor. EPC contractors will usually resist such a condition on the basis that it can be used to coerce the contractor into an unfair settlement of an equitable adjustment claim in its desire to achieve substantial completion or provisional acceptance and be paid its final milestone payment.

Substantial completion or provisional acceptance will mark the reversion of care, custody and control (discussed in Chapter 12) of the facility from the EPC contractor back to the owner and will usually trigger commencement of the operating insurance coverage for the facility. This is another reason to bar equitable adjustment claims after the date of substantial completion or provisional acceptance. No work except punch list items will be performed after substantial completion or provisional acceptance, and the owner has risk of loss anyway at this time. Thus, there should be nothing for which the EPC contractor would need an equitable adjustment.

In summary, "substantial completion" and "provisional acceptance" are not construction concepts. They are commercial concepts and their list of prerequisites should not be relegated to one of the engineering exhibits of the EPC contract to be settled by technical professionals working in isolation. Everyone involved in a project should review the list of conditions for the achievement of substantial completion or provisional acceptance before it is included in the EPC contract. As a legal drafting matter, it is probably preferable for the owner to use the term "provisional acceptance" rather than "substantial completion" because "provisional acceptance" seems to imply acceptance by the owner is still subject to conditions, which seems to create an advantage for the owner over the term "substantial completion" (which seems to imply the facility is complete) if the owner discovers issues later. In either case, though, there should not be much difference in the end if the EPC contract has been properly drafted.

Final Acceptance

Since it is not practical or necessary to finish all facility-related work by the date of substantial completion or provisional acceptance, usually the EPC contract will specifically include a list of items that need not be accomplished for the EPC contractor to achieve substantial completion or provisional acceptance (such as painting, paving and landscaping). Also, to the extent the owner agrees, a punch list of items will often be left to be completed by the EPC contractor after substantial completion or provisional acceptance occurs. "Final acceptance" or "project closeout" or "project completion" will occur under the terms of the EPC contract when no work (including any punch list work) remains to be done and all deliverables have been delivered as required by the EPC contract.

Early Operation

Situations can arise in which the owner will want to operate its facility before the EPC contractor has achieved substantial completion or provisional acceptance. Accordingly, the EPC contract should include provisions to provide for "early operation." An EPC contractor will generally resist inclusion of early operation provisions.

There are two common cases in which it may be desirable for the owner to take over the facility before the EPC contractor has achieved substantial completion or provisional acceptance. The first, and less compelling case, is the situation in which the owner stands to lose a profit if it does not operate its facility as soon as possible. The second, and more critical case, is the situation in which the owner stands to lose its offtake, use or concession agreement altogether if the facility is not placed into service by the in-service deadline required in those agreements.

In the first instance, the facility might be a plant being developed as a "merchant" power plant that will sell its electrical output into a power pool on a spot basis. Or, in the case of a power plant that does have an offtake agreement, the owner may have reserved the right to sell power into a power pool before the plant passes the offtaker's tests for being placed into service under the offtake agreement. In all of these cases, revenue foregone can never be recovered. While the owner may not have taken this revenue into account when calculating whether or not to construct the project, there is little reason that the owner should not try to harvest some of this revenue if its facility is operable. For example, consider the circumstances in which the EPC contractor is scheduled to complete a three unit merchant power plant in June so that the power plant will be in operation for the summer peak electrical demand. Suppose the EPC contractor successfully completes one of the three units by June but the other two will likely not be ready until August or September. If the owner waits until August or September to accept the entire power plant, even though the owner may receive "delay" liquidated damages from the EPC contractor for the delay in commercial operation, it may be that the owner would earn more money by putting the one complete unit into operation in June. In fact, larger facilities are often placed into service in stages for just this reason. In addition, the EPC contractor probably no longer wants to bear the responsibility of the risk of loss (discussed in Chapter 12) for the completed unit because accidents can happen. Therefore, the EPC contractor often desires to transfer care, custody and control of different components of a facility to the owner as soon as possible.

A more dire situation is the one in which the EPC contractor will be late completing the facility and there is some question about whether the offtaker or user will grant the owner an extension of the date under the offtake or use agreement by which the facility must be put into service. Thus, if the facility is capable of safe operation but not yet capable of meeting the performance promised by the EPC contractor in the EPC contract, the owner may want to take over the facility and place it into service despite the fact that the EPC contractor is not yet finished. As discussed above, offtakers usually have a battery of tests to be performed to demonstrate that a facility can operate reliably. These tests are usually much less stringent than the tests imposed on the EPC contractor under the EPC contract. Therefore, it is conceivable that the facility could pass the offtaker's tests (which generally relate primarily to safety and output level) although the facility is not ready for all the tests called for by the EPC contract (particularly those that relate to the facility's efficiency). In the case of an offtaker that is just buying a facility's output, the offtaker is not likely to have a great concern about efficiency because the offtaker generally will not be paying for the facility's feedstocks and fuel. However, in the case that involves a so-called "tolling" or "conversion" agreement in which the "toller" is paying a fee to the owner so that the owner can convert raw materials or fuel into another product, the toller will be very concerned about the conversion efficiency of the plant because the plant will require more feedstocks or fuel than expected. This is why, in a tolling agreement, an owner usually has to guarantee a conversion efficiency to the toller and pay liquidated damages if the plant does not achieve that efficiency.

If the owner has financed its project, another issue could be that the owner's construction period loans are about to mature and the owner must demonstrate that its facility is operable so the lenders will honor their commitment to extend the maturity of the construction period loans into term loans. However, the test for commercial operation imposed by lenders usually requires that the facility pass all performance tests imposed by the EPC contract. Thus, the owner may not be able satisfy its lenders' requirements for commercial operations until the EPC contractor can achieve the performance required by the EPC contract. If this is the case, the owner should determine whether it and the EPC contractor can (without being required to obtain the lenders' consent under the loan agreement) modify the performance tests under the EPC contract so that the facility will pass the EPC contract tests.

EPC contractors will usually insist that, if the owner puts the facility into "early" operation, the facility will be deemed to have passed all the performance tests and care, custody and control and risk of loss of the facility will transfer to the owner. To support its logic, the EPC contractor will contend that the owner must be satisfied with the facility if the owner is willing to place the facility into operation. This may not, as was noted in the cases above, be true. A fair compromise of the owner's right to place the facility into "early" operation is usually to provide that the facility not be deemed "substantially complete" or "provisionally accepted" and still must pass performance tests set forth in the EPC contract once the EPC contractor finishes its work. However, to the extent that the owner's "early" operation of the facility causes the EPC contractor's work to be delayed, or more costly, the owner will compensate the EPC contractor for these consequences through an equitable adjustment and corresponding change order.

Also, typically the warranty period under the EPC contract commences if the facility is placed into "early" operation as mentioned above because the warranty covers operation of the facility and the facility has, in fact, been placed into operation. Whether care, custody and control should transfer to the owner upon "early" operation should be examined in conjunction with the insurance program for the project. Generally, the EPC contractor will be eager to transfer care, custody and control to the owner, but the owner should be careful about accepting responsibility for an incomplete facility and should confirm provisions of its insurance policies to determine that the builder's "all-risk" insurance covering the work will not terminate if the owner assumes care, custody and control of the facility (see Chapter 23). During "early" operation, the EPC contractor is usually excused from paying schedule delay damages because the EPC contractor will usually not have unfettered access to all areas of the facility for purposes of construction and testing, therefore, it is not reasonable to penalize the EPC contractor for this lack of access.

Operating Revenues during the "Early" Operation Period

One way to make the EPC contractor more receptive to the possibility of "early" operation is to share some portion of the revenue or profit produced during this period with the contractor. However, this is rarely done because it puts the EPC contractor in the position of being the *de facto* partner of the owner and possibly creates the wrong incentive for the EPC contractor in that the EPC contractor's tardiness in performance has the potential of rewarding it with a "piece of the action" rather than giving rise to penalties in the form of delay liquidated damages.

In addition, "early" operation as profit sharing is difficult to administer in practice. For example, in the case of a merchant power plant, who should decide when the power plant will run? The owner or the EPC contractor? How will profit be calculated? Before or after fixed costs are paid? Truthfully, a provision for "early" operation itself is likely never to be read again once the EPC contract has been executed, but in the event that it is needed, the owner may deeply regret its absence.

Notes

1 The concept of "equivalent operating hours" has developed in order to take into account the fact that certain activities place additional stress on a plant's components and therefore such activities should be accounted for as more than just normal operating hours. For example, turbines usually perform best when they are running continually rather than starting and stopping. (In fact, steam turbines in nuclear power plants are usually designed to run for years without stopping.) Thus, startup and shutdown (especially if these activities are performed more quickly than the manufacturer's recommended "ramp up" and "ramp down" times) and excessive or inadequate electrical demand from the power grid, can all take their toll on equipment. An hour in "startup" mode may actually equate to three hours of "normal" operation. If a turbine is designed to be started once per day and it is stopped and started four times during each day of the year, it will probably degrade much more quickly than a turbine that has been started only once each day of the year even though the overall operating hours of the turbines in these two cases may be very similar. In order

to account for different regimes of operation, the notion of "equivalent operating hours" is used.

2 The higher the power factor, the higher the active electrical output of the power plant. See note 4 in this chapter.

3 A note about power plant nomenclature may be useful. "Full load" and "name plate" or "nominal" rating are not necessarily the same. A power plant that has a "name plate" or "nominal" rating of 500 MW usually means that the power plant has been designed to generate 500 MW at ISO conditions of 59°F, 60 percent relative humidity, and atmospheric pressure of 1.013 bar. At "full load" this power plant might or might not produce 500 MW. "Full load" in a natural gas-fired plant refers to the condition that the throttle is wide open and allowing the maximum amount fuel to flow into the combustor. If the ambient air is cold and humidity is low, the power plant is likely to produce more than 500 MW when the fuel throttle is completely open. If the ambient air is hot and it is humid, the power plant is likely to produce less than 500 MW when the fuel throttle is completely open.

4 Generators typically produce some mixture of "real" and "reactive" power. The balance between them can be adjusted on short notice to meet changing conditions. Real power, measured in watts, is the form of electricity that powers equipment. Reactive power, a characteristic of alternating current systems, is measured in volt-amperes reactive, and is the energy supplied to create or be stored in electrical or magnetic fields in and around electrical equipment. Reactive power is particularly important for equipment that relies on magnetic fields for the production of induced electric currents (motors, transformers, pumps and air conditioning, for example). Transmission lines both consume and produce reactive power. At light loads, they are net producers. At heavy loads they are heavy consumers. Reactive power consumption by these facilities or devices tends to depress transmission voltage, while its production (by generators) or injection (from storage devices such as capacitors) tends to support voltage. Reactive power can be transmitted over only relatively short distances during heavy load conditions. If reactive power cannot be supplied promptly and in sufficient quantity, voltages decay. In extreme cases a "voltage collapse" may result.

Power grids set two types of stability limits. First, voltage stability limits are set to ensure that the unplanned loss of a line or generator (which may be providing locally critical reactive power support) will not cause voltages to fall to dangerously low levels. If voltage falls too low it begins to collapse uncontrollably, at which point automatic relays either shed load or trip generators to avoid damage. Second, power (angle) stability limits are set to ensure that a short circuit or an unplanned loss of a line, transformer or generators will not cause the remaining generators and loads being served to lose synchronism with one another because all generators and loads within an interconnection must operate at, or very near, a common frequency. Loss of synchronism with the common frequency means generators are operating out-of-step with one another. Even modest losses of synchronism can result in damage to generation equipment (source: USCBR).

5 While the owner may have some defenses (such as "laches" or "estoppel," which can bar claims that have not been asserted with reasonable promptness) against the EPC contractor raising unreasonably late claims, the inclusion of an absolute cut-off date in the EPC contract is advisable. See *Bulley & Andrews, Inc. v. Symons Corp.*, 323 N.E.2d 806 (Ill. App. 1975), holding that a contractor's objection to a method of using threaded tie rods to hold cement forms together (instead of looped tie rods) that was lodged nine months after substantial completion, was precluded as a result of the contractor's silence at the time the method was imposed by the owner.

Chapter 17

Liquidated Damages for Delay and Impaired Performance

As was discussed in Chapter 13, once a party can demonstrate that it is entitled to damages as a result of the other party's breach of a contract, the party must demonstrate the precise amount of its damages, and the breaching party will have the opportunity to dispute the proffered damage amount. The liable party may be able to reduce the amount of damages that it must pay, either by demonstrating that the other party's actual damages are not as extensive as claimed, or by showing that the other party could have mitigated the amount of its damages by taking precautionary or remedial measures (as was discussed in Chapter 13).

In cases in which it is relatively easy to demonstrate a party's benefit of its bargain and calculate its cost to cover the loss of its bargain, there is usually little argument. If a coal supplier fails to deliver coal at $40/ton and the buyer must purchase replacement coal for $41/ton, it is easy to see the extent of the buyer's economic injury. But what about the movie production company that signs a contract with a leading actor and then must replace him when he fails to show up for filming? How can the producer "prove" the difference in what the box office receipts might have been had the leading actor starred in the movie?

In order to address situations in which the level of damages is more speculative, the construct of "liquidated damages" developed. When it would be difficult or impossible to calculate a party's harm with precision, courts permit parties to employ the concept of a liquidated amount as damages in lieu of actual damages so long as the liquidated amounts are a reasonable estimation of the damage that would have been incurred as the result of a breach of a party's obligations and not an amount being used as a penalty to deter a party from breaching its contract (contractual penalties are void in the United States and the United Kingdom).[1] Also, the payment of penalties is generally not covered by most insurance policies so EPC contractors should be sure not to permit the use of the word "penalty" as a synonym for "liquidated damages." Otherwise, the EPC contractor may not be covered under its delayed startup insurance for any delay liquidated damages that it must pay (discussed in Chapter 23).

Although it is difficult to find a court decision discussing this principle, courts are said not to discourage "efficient" contractual breaches of commercial contracts—that is, the case in which a party chooses to breach its contract because it can afford to pay for the consequences of its actions. In the previous example, suppose the coal supplier could sell its $40/ton coal for $44/ton to another buyer and the original $40/ton buyer must buy coal for $41/ton if the coal supplier does not honor its delivery obligations to the buyer. The supplier could pay the original \$40/ton buyer for its \$1/ton loss and the supplier could still earn a \$3 profit on its sales to the \$44/ton buyer. As the theory goes, no one is harmed in the above case. Conversely, U.K. courts are reluctant to enforce clauses whose predominant purpose appears to be the defense of contractual breach.[2]

This hypothesis may be valid for commodities like gold, copper, iron ore, grain and gasoline, but it has its limitations in other contexts. In fact, even in the case of coal, output from different mines has different energy and sulfur content. A facility's boilers may be designed to burn Powder River Basin coal, while coal from another region like West Virginia may pose operating or emissions problems for the facility. Thus, it may not be so simple for a court to determine how to compensate a buyer of coal for its losses. Compensation for a duped buyer is especially difficult in some transactions, such as real estate, which is why courts may be inclined to use their equitable powers to award specific performance in real property transactions (as was discussed in Chapter 13).

Courts also generally require that the contract parties had equal bargaining power when entering into the contract in order to enforce a liquidated damages provision. The contract must not be what is known as an "adhesion" contract in which one party has no power to bargain with the other (such as the purchase of a consumer item like a toaster). In the case of infrastructure facilities such as power plants, it is difficult to calculate precisely what damages the owner will suffer if the EPC contractor fails to perform, but it is possible to make an estimate of the owner's profit by using reasonable assumptions for interest rates, feedstock prices, fuel costs and operation and maintenance costs (see Table 4.1 in Chapter 4). Thus, liquidated damages are a useful mechanism for compensating the owner for the EPC contractor's delayed or poor performance. However, it is important for the EPC contract to provide that actual damages (instead of liquidated damages) can be recovered by the owner if a court will not enforce liquidated damages provisions of the EPC contract.[3]

Liquidated Damages for the EPC Contractor's Failure to Achieve Substantial Completion or Provisional Acceptance on Time

Under the EPC contract, the EPC contractor will agree[4] that substantial completion or provisional acceptance will be achieved by a specified date. This is commonly referred to as the "guaranteed completion date" or the "guaranteed commercial operation date" or the "guaranteed COD date." What happens if the EPC contractor does not achieve substantial completion or provisional acceptance for the facility by the guaranteed date? The EPC contractor will pay liquidated damages to the owner for this delay. These are often referred to as "schedule lds" or "delay lds" or "schedule delay damages."

Delay liquidated damages are payable to the owner in recognition of the fact that once the opportunity to produce and sell output on any day is lost, the owner will never again be able to earn any revenue for that day, although the owner will have incurred interest charges on its loans outstanding on that day and possibly also have had to pay delay liquidated damages to its offtaker or user as well (because the offtake or use agreement generally contains delay liquidated damages provisions for the owner's failure to achieve commercial operation by the date the owner has "guaranteed" that the facility will be ready for commercial operation). In such cases, the level of the EPC contractor's delay liquidated damages will usually be set to cover the owner's interest costs and the delay liquidated damages that the owner must pay to the offtaker. Clearly, lenders will not want a situation to arise in which the owner has no money to pay delay liquidated damages to its offtaker or user because this could lead to the owner's payment default under the offtake or use agreement, which often will give the offtaker or user the right to terminate the agreement and leave the owner and lenders with a "stranded" facility that may or may not have (for regulatory or physical reasons) a market in which to sell its output. Even if a market exists, market prices may not be enough to cover the owner's costs.

For simplicity's sake, liquidated damages for delay usually accrue on a daily basis. The owner's daily profit depends on fuel costs and the owner's interest costs. Obviously, profit can vary significantly if fuel prices change (unless the owner has hedged its fuel risk, which can be done either financially, with a derivative instrument, or physically, with a fuel storage arrangement). Rarely, if ever, will the delay damages cover the owner's expected daily profit. The EPC contractor will usually not agree to such a high level of liquidated damages because the EPC contractor's profit could be wiped out quickly by a significant delay. Also, it is not really in the owner's negotiating interest to disclose its expected profit or even its total project capital cost to the EPC contractor because the EPC contractor might then try to estimate the owner's profit. If the EPC contractor believes that the owner's profit margin is high, the EPC contractor may try to extract some of this margin for itself, either by raising the EPC contract price or being aggressive about the pricing of change orders if unexpected circumstances arise. An EPC contractor usually will ask what the owner expects its capital cost of the project will be. Simply put, this is none of the EPC contractor's business.

Delay liquidated damages are often scaled to increase as the EPC contractor's delay in completing the facility continues, in recognition of the fact that the owner's lost profits mount and often cannot be recovered unless the offtake or use agreement for the facility is extended by the offtaker or user (even if the offtake or use agreement is extended by the offtaker or user, the owner's profits in this future period are probably speculative at best because this period may be very far in the future and the condition of the facility and economic conditions may change dramatically).

For facilities comprised of trains that can operate individually, it is usual to reduce the EPC contractor's delay liquidated damage payments if one or more trains can be placed into operation, so long as the offtaker or user (if there is one) will accept a reduced level of output temporarily (or even permanently). Finally, it is important that the EPC contract make clear when delay liquidated damages can be invoiced by the owner and when they must be paid by the EPC contractor, for example, daily, weekly, monthly, or at the end of the delay.

It is also important to make sure that the "guaranteed" completion date should not change for any reason unless the EPC contractor is entitled to an equitable adjustment under the EPC contract because one of the circumstances specifically enumerated in the EPC contract (such as *force majeure* or owner delay) has triggered an extension. As was discussed in Chapter 14, clauses in the EPC contract that attempt to postpone the "guaranteed" substantial completion date (such as a clause that states that "unless not due to the EPC contractor's fault, completion will be achieved by the guaranteed date") should be avoided. Why? Because the EPC contract's equitable adjustment mechanism is designed to handle and specifically identify the only instances that entitle the EPC contractor to a postponement of the "guaranteed" completion date. Moreover, "fault" is not really a legal concept or theory. It is really a concept used by insurers. A plaintiff in a court case proves the "negligence," not "fault," of a defendant. The inclusion of fault language in an EPC contract has the potential to undermine the reason why the owner entered into an EPC contract and agreed to pay the EPC contractor the risk premium that a general contractor would not have commanded. Fundamentally, if a project does not achieve completion on time, it is the EPC contractor's responsibility (and fault) unless the EPC contractor can demonstrate by a "preponderance of the evidence" (the evidentiary standard in civil cases as opposed to "beyond a reasonable doubt," which is the evidentiary standard in criminal cases) that an event entitling the EPC contractor to an equitable adjustment has occurred. In fact, in negligence cases, typically the doctrine of *res ipsa loquitur* (Latin for "the thing speaks for itself") operates to shift the burden of proof of negligence from the plaintiff to the defendant.[5] Including fault language in the EPC contract could have the catastrophic effect of shifting the burden of proof from the EPC contractor's being required to show that it is entitled to an equitable adjustment to the owner's being required to demonstrate that the EPC contractor was at "fault." Suppose, for example, in the case of a power plant, that the EPC contractor has chosen a responsible railroad carrier to transport a turbine but the carrier's flatbed rail car derails and the turbine is damaged. Is this the EPC contractor's "fault"? Will the EPC contractor be able to demonstrate that it was not at fault for this event? Will the owner be able to prove that this event was the EPC contractor's fault? From an owner's point of view, it seems far better for the EPC contractor to be compelled to claim that this was an event beyond the EPC contractor's control, which constitutes *force majeure* (see Chapter 12) and entitles the EPC contractor to an equitable adjustment rather than the owner trying to convince an arbitrator, judge or jury that this event was the EPC contractor's fault.

As will be discussed later in this chapter, it is unlikely that the EPC contractor will agree to assume unlimited liability for delay liquidated damages and will insist upon an agreed limit on its payment of liquidated damages for delay. This is reasonable but the owner should always be certain that the EPC contract makes clear that, while the EPC contractor's liability with respect to late completion of the facility is limited, the EPC contractor is obligated to complete the facility itself in all circumstances whether or not it has reached a payment limit on delay liquidated damages. In fact, the EPC contract must contain an express "event of default" provision that provides that the EPC contractor will be in default under the EPC contract if the EPC contractor does not complete the facility within a certain number of days after the "guaranteed" completion date. This date will usually be the date on which the EPC contractor's obligation to pay delay liquidated damages is exhausted because the payment limit has been reached. Once this default occurs, the owner will be free to pursue its remedies (discussed in Chapter 13), if necessary. For the owner, this "event of default" provision is probably the single most important event of default to include in the EPC contract because, if completion of the facility is so significantly delayed, the owner needs to have leverage to compel the EPC contractor's performance. The threat of termination of the EPC contract may elicit the EPC contractor's expeditious action because the EPC contractor is likely to lose any profit it hoped to make if the owner terminates the EPC contract (and then, in addition, the owner may seek to charge the EPC contract to recover the cost of any replacement EPC contractor that the owner retains). On the other hand, there may be no profit at all left in the project for the EPC contractor at this point and thus the EPC contractor may be considerably less motivated to complete the facility. While the EPC contractor's liability for delay in completion of the facility is limited to an agreed-upon level of liquidated damages, the EPC contractor's liability for actually completing the facility (late or not) is usually only limited to the total EPC contract price (as will be discussed in Chapter 20).

Early Completion Bonuses

If it is in the owner's interest to commence commercial operation as soon as possible, occasionally the owner will (as was mentioned in Chapter 16) offer a monetary incentive to the EPC contractor to achieve completion before the "guaranteed" completion date. This bonus can take the form of a liquidated bonus payment or a share in the profits or revenues to be earned from the facility during this early operating period. If this is so, the EPC contractor should insist on a share of revenue receipts as opposed to a "share" of the owner's profit. As can be learned from the movie business, the box office "take" for a film can be high but there very well may be no profits left after expenses and royalties have been charged. An EPC contractor should not have to sort out the (possibly recondite) accounting practices of the owner. In reality, owners tend to be reluctant to offer bonuses, perhaps because many owners feel that EPC contractors are often in enough of a rush to complete projects without the incentive of a bonus. Offering a bonus could also make it more difficult for the owner to understand the true project schedule. It seems more likely that the EPC contractor will be inclined to select a "guaranteed" date that is very easily achievable so that the EPC contractor has a realistic chance of completing the facility early and earn a bonus. If a bonus is offered, owners often do not permit the "bonus" date to change (even as a result of *force majeure*). The guaranteed completion date can change, of course, as a result of an equitable adjustment, but owners often do not permit the "bonus" date to change. Many owners believe that not permitting the "bonus" date to change for any reason creates the proper incentive for the EPC contractor.

Liquidated Damages for Impaired Performance

What happens if a facility does not perform as well as expected when it is completed? Obviously, if it performs below a certain level, it is of little or even no use to the owner. What about the case in which the facility's performance is acceptable but not ideal? Typically, if the facility has reached a threshold level of performance, the owner will agree to accept it so long as the EPC contractor pays liquidated damages to compensate the owner for the consequences of this below-anticipated performance level. Usually, the owner will acknowledge that the facility is acceptable only once certain agreed-upon minimum levels of performance have been achieved. Often, the EPC contractor is given a cure period (either before or after it pays liquidated damages to the owner for not meeting the targeted performance) within which the

EPC contractor must make efforts to improve the facility's performance so that the funding will meet the target levels "guaranteed" by the EPC contractor in the EPC contract. If the EPC contractor is given a cure period, the EPC contract should specify whether the owner will be able to operate the facility during this period (which is typical) and whether or not the EPC contractor will have the ability to interrupt the owner's commercial operations during this period. Should an EPC contractor be given this right, it is important that the EPC contractor be required to give the owner advance notice of when the work will impede or prevent the facility's operation. If the EPC contract does not contain provisions explaining how the owner and the EPC contractor are expected to coordinate during this period, the owner may be exposing itself to an equitable adjustment claim from the EPC contractor.

Output

Generally, schedule delay damages cease to accrue (sometimes only temporarily) once the EPC contractor achieves the minimum performance level set forth in the EPC contract. Typically, in the case of a power plant, these minimum levels are 95 percent of the EPC contractor's electrical output target level and no more than 105 percent of the EPC contractor's efficiency (heat rate) target level. If, by the end of the cure period, the facility can meet the performance guarantees contained in the EPC contract, no performance liquidated damages will be due to the owner or the performance liquidated damages will be refunded by the owner (or released from escrow) if they were paid to the owner (or to an escrow agent). If, however, the facility cannot achieve these target levels, the EPC contractor must pay liquidated damages to the owner to "buy down" the levels promised by the EPC contractor in the EPC contract. That is why liquidated damages for poor performance are often referred to as "buy down" damages.

If an EPC contractor is having difficulty in achieving the target level of facility performance, the EPC contractor will often calculate whether it will be more expensive for the EPC contractor to pay delay liquidated damages while the EPC contractor works to achieve better facility performance or whether it should just commence facility performance tests and pay performance liquidated damages to the owner for the inferior performance of the facility. Furthermore, in the case of the payment of performance liquidated damages, the EPC contractor may be entitled to reimbursement from a vendor for a substantial portion of the performance liquidated damages that it will pay to the owner if the performance problem is caused by vendor-supplied equipment. In fact, if the owner has chosen and purchased some of the main equipment for the facility directly from a vendor (such as the turbine generator in the case of a power plant), the EPC contractor will insist on reducing the amount of performance liquidated damages that will be payable to the owner to the level of performance liquidated damages payable by the vendor to the owner with regard to its equipment if the EPC contractor can demonstrate that such vendor's equipment is causing the shortfall in performance of the facility. The EPC contractor's premise is that it was the owner that negotiated that level of performance liquidated damages payable by the vendor for its equipment and the EPC contractor should not pay a penalty on top of that amount if the EPC contractor has otherwise built the facility without any other deficiencies in performance.

For example, suppose the EPC contractor has promised a two-unit 100 MW facility will be ready in 24 months but is already in month 22 and preliminary tests show that the power plant is capable of generating only 94 MW because the parasitic load is greater than anticipated. Suppose the EPC contractor believes it would take one month to make changes to the equipment to reduce the parasitic load to 5 MW to meet at least the 95 percent minimum level of the 100 MW it promised and thereby achieve completion by the "guaranteed" completion date so that no schedule delay damages will have to be paid (but the EPC contractor will have to pay performance liquidated damages for the 5 MW shortfall because the power plant will only achieve a 95 MW output during its performance tests [and not 100MW]). On the other hand, what if the EPC contractor believes that electrical design modifications taking two to three months could be made and implemented to achieve the full 100 MW performance guarantee for electrical output (of course, during the "correction" period, the EPC contractor will have to pay delay liquidated damages to the owner since the power plant would not have achieved substantial completion). In this case, the EPC contractor will evaluate its options of trying to bring the output up to 96 MW and paying performance liquidated damages for the 4 MW electrical output shortfall or taking the time necessary to bring the power plant up to the full 100 MW performance while paying delay liquidated damages to the owner since the power plant will not have achieved substantial completion because it will be incapable of producing even 95 MW while modifications are undertaken. On the owner's side of the equation, the owner would probably rather wait two or three months for the extra 4 MW because the owner will probably make more money selling 100 MW than 96 MW during the lifecycle of the facility. However, if the owner has a sunset date for achieving commercial operation with its power customer and there is a significant chance that the EPC contractor's re-design and rebuilding could be unsuccessful and, therefore, the power plant may not meet the 95 MW minimum level by the sunset date, the owner could lose its power purchase agreement with its customer. So, taking this risk in order to gain 4 MW of capacity might not be acceptable to the owner. If the owner has financed its project, the owner's lenders will be even less interested in taking this 4 MW risk, especially since any performance liquidated damages received by the owner will be required to be paid by the owner to the lenders as prepayments of their loans. The lenders will require this because, when they lent their money, the amount of their loans was premised on the assumption that the power plant would have a certain amount of installed electrical output (capacity) and a certain efficiency (heat rate) so that sufficient revenues could be generated to repay their loans. At a reduced output or efficiency level, revenues and/or profits will be reduced. Therefore the lenders have more risk and their debt service coverage ratio (discussed in Chapter 4) will also be reduced. Thus, in a sense, the lenders have lent the owner too much money and must collect the performance liquidated damage payments the owner receives from the EPC contractor to correct this "over extension" of credit to the owner. Lenders to a project generally are willing to expose themselves to "operating" risk (the risk that the owner is not a good operator) and therefore the owner may not earn enough money to repay the loans on the required amortization schedule (or perhaps in full ever) and sometimes are prepared to expose themselves to commodity risk (the risk that input, fuel or output prices will change) and therefore they will have the same repayment issues as in the case in which the owner is a poor operator but they will usually not be prepared to expose themselves to the risk of impaired performance because the facility has not been built to the minimum specification set forth in the EPC contract. If the owner has included a properly drafted "early" operation provision (as discussed in Chapter 16) in the EPC contract, the owner will not have to stand by at the EPC contractor's mercy while the EPC contractor decides which course of action it will take (that is, fix the problem or abandon remedial measures). Under an "early" operation provision the owner could take control of the facility and operate it.

EPC contracts for processing facilities such as air separation units will often contain what is called a "make right" or "put right" obligation for output and efficiency. This means the owner will accept no lesser performance for the facility than the levels promised by the EPC contractor in the EPC contract. This can be the case because certain levels of output and efficiency are necessary to ensure that the overall process is efficient.

Efficiency

Inferior efficiency of a facility is usually more serious than inferior output from the facility. A poorer efficiency than anticipated means that the facility will consume more fuel or raw materials forever. Thus, an increase in fuel or raw material prices (as has been true for oil and gas in the past) can be potentially lethal to a project's profitability.

Availability

As has been discussed, it is important for an owner to know what type of availability can be expected from its new facility. Many things (including sunspots in the case of large transformers) can affect a facility's availability and EPC contractors are not insurers of the availability of the facilities they build. While it is not common, EPC contractors and major equipment vendors sometimes do, however, offer "availability guarantees" to owners that "guarantee" that the facility (or certain of its equipment) will, after a predetermined shakeout period, achieve a certain level of availability or the EPC contractor (or vendor) will pay liquidated damages to the owner for the facility's failure to attain the target level of availability. These "guarantees" are often offered when a vendor is trying to market new technology (such as direct drive electric motors) or an EPC contractor is trying to move into a new market segment (such as windmills or integrated gasification in the case of power plants). Availability guarantees may have the consequence of raising the EPC contract price (because the EPC contractor has taken on more risk). Also, EPC contractors and vendors that do give availability guarantees often impose operating restrictions on the owner so they can ensure that the facility is operated in accordance with all instructions and manuals so that availability will not suffer unnecessarily. (In fact, they may even insist that they be the operator of the facility.) The EPC contractor may also want to monitor facility performance on a real-time basis by direct "hook in" to the facility's control room. Sometimes the EPC contractor will even post an observer on site. Availability guarantees are most common when the owner has also entered into a long-term maintenance contract with the EPC contractor or vendor regarding major equipment of the facility.

Emissions

Emissions levels are also "guaranteed" by the EPC contractor in the EPC contract. These "guaranteed" levels are typically not subject to relaxation through "buy down" liquidated damages because facilities must operate within their permit requirements or face fines (or even closure and criminal prosecution).

Emissions into the atmosphere can be controlled in basically three ways with varying success. First, the composition of fuel and raw materials can be controlled (for example, using coal with a low sulfur content). Second, the combustion process can be made more efficient by converting as much fuel as possible into energy (for example, using "reheat" burners that fire sequentially). Third, emissions can be eliminated before their release into the atmosphere (for example, by use of electrostatic precipitators, which use an excited electromagnetic field to attract particulate matter before it can escape).

Other emissions from a facility can also be noxious. It will be the EPC contractor's responsibility to meet the owner's requirements in these cases as well. Water discharge from a facility has the potential to upset aquatic ecosystems if its temperature is too high or too low. In addition, excessive discharges of elements such as phosphorus and nitrogen can nourish plant growth and lead to eutrophication in which excessive plant growth (such as algae) leads to decreased oxygen levels in water because bacteria use oxygen to break down plant growth.

Excessive water intake levels also have the potential to deplete groundwater tables or affect the salinity of water basins if an insufficient amount of water is returned to the source from which it has been taken.

Noise and electromagnetic radiation levels also can be of issue if facilities are located in populated areas.

"Double Counting"

If the owner has imposed performance liquidated damages on not only the entire facility but also its individual units, the EPC contract should not penalize the EPC contractor twice for the same problem.

Suppose, for instance, an EPC contract requires that each of the two units in a power plant achieve an output of 50 MW. An owner might insist on this to make sure that if one unit unexpectedly trips or is taken out of service for maintenance, the other unit will still be available to sell 50 MW (and not some lesser amount) to its offtaker or into the market. Suppose one unit is performing at an output of 46 MW and the other is performing at 54 MW? Should the owner be happy because the power plant will be able to generate the 100 MW the owner was expecting? It depends on what type of service is expected for the plant. If the owner has required that minimum performance levels be attained for each of the units, the EPC contractor is likely to have requested that superior performance on one unit can be used to offset inferior performance on the other unit.

Owners may agree to the offsetting of performance deficiencies but will generally limit the amount of the benefit the EPC contractor can allocate from one unit to another. First, the owner wants two reasonably similar units and does not want a serious output or efficiency problem if the superior unit turns out to be operationally unreliable. Second, if the unit can easily exceed its performance guarantees, it may be that the guarantees were set at levels too low for the equipment the EPC contractor is actually supplying and, therefore, the owner may be paying for a larger machine than is actually needed. Thus, the owner may agree to let the EPC contractor allocate up to 2 MW from one unit to the other unit. In the example, performance of the 46 MW unit would be raised to 48 MW and the EPC contractor would pay liquidated damages on the 2 MW shortfall on the substandard unit between 48 MW and the 50 MW guarantee but no liquidated damages on the entire facility would be assessed because the facility can perform at the required 100 MW when both units are in operation.

Alternatively, consider the power plant efficiency (heat rate) in which the guarantee is a maximum of 7,150 MMBtu/kWh for each unit and a maximum of 7,100 MMBtu/kWh for the facility. Suppose further that during performance tests it is revealed that the heat rate for the units is 7,200 MMBtu/kWh and 7,190 MMBtu/kWh, respectively, obligating the EPC contractor to pay performance liquidated damages for each of the units because neither has met the 7,150 MMBtu/kWh guarantee. If both units fail to achieve their heat rate guarantees, it is highly improbable that the facility can achieve its heat rate guarantee. If the owner charged the EPC contractor liquidated damages for the facility's failure to meet its heat rate guarantee in addition to charging the EPC contractor liquidated damages for the individual units' failure to attain their guaranteed performance levels, the EPC contractor would, in essence, be paying the owner twice for the same problem.

In addition to violating basic tenets of fairness in commercial dealings, charging the EPC contractor twice for the same problem seems to be penal and in the United States the owner could run the risk of a court invalidating the liquidated damage provisions of an EPC contract entirely as punitive and against public policy. In such a case, the owner would have to prove its actual damages and, since (as was discussed above in this chapter) courts will not award "speculative" damages (such as those based upon future fuel and power prices), the owner might have difficulty recovering damages that are being sought.[6] Especially in the case of a heat rate problem, the level of damages is highly dependent upon the future price of fuel over the life of the power plant. Proving future fuel prices could be an evidentiary nightmare for an owner. The owner is unlikely to recover anything unless it can prove these levels by a preponderance of the evidence, which may be quite difficult because the EPC contractor is likely to retain fuel market experts who will dispute the owner's posited price levels.

All these issues can likely be avoided if the EPC contract has a mechanism to prevent such double counting from occurring. If, however, the court, for some reason, rules that the liquidated damage provisions of the EPC contract are unenforceable or void, the owner still may have recourse against the EPC contractor so long as the owner can prove its damages. Typically, in another section of the EPC contract, the EPC contractor will have made a representation to the effect that the EPC contract was validly authorized and duly executed by the EPC contractor and is enforceable against the EPC contractor in accordance with its terms. If a court fails to uphold the liquidated damage provisions of the EPC contract, the owner will be able to make a claim against the EPC contractor for breach of this representation made by the EPC contractor if the representation turns out not to be true because the EPC contract's liquidated damages provisions are held by the court to be unenforceable (see Chapter 22).

"Caps" on Liquidated Damages for Delay and Poor Performance

Since problems can arise and their consequences will be unknown at the time of execution of the EPC contract, to guard against grave situations that could require the EPC contractor to spend far more money in the payment of liquidated damages than the EPC contractor's expected profit on the EPC contract, it is standard for EPC contractors to limit their liability to pay liquidated damages for delay in completing the facility and diminished performance of the facility to some percentage of the EPC contract price (typically between 10 and 35 percent). The cap will usually depend upon the EPC contractor's reputation and the size, complexity and location of the project. After these limits of liability are reached, the EPC contractor must still finish the facility (and the EPC contract should expressly require this) but the EPC contractor will no longer be responsible for paying the owner liquidated damages for delays in completing the facility and performance level shortfalls of the facility.

"Subcaps"

In an attempt to further reduce their exposure, most EPC contractors will request "subcaps" on their liability for the payment of liquidated damages. Thus, in an EPC contract that contains a 25 percent limit on liquidated damages, the EPC contractor may request a 15 percent "subcap" for delay damages. This means that if the facility is delayed, the EPC contractor would pay only up to the 15 percent subcap on delay liquidated damages instead of 25 percent and 10 percent of the EPC contract price would still be left for payment of liquidated damages in the case that the facility also suffers from reduced performance once it is complete.

Notes

1 See *Dunlop Pneumatic Tyre Co Ltd v. Selfridge & Co Ltd* [1915] UKHL 1.
2 *M&J Polymers Ltd v. Imerys Minerals Ltd* [2008] EWHC (Comm) 344, [2008] 1 All ER 893 (EWHC) (Comm) (Eng.).
3 See *Truck Rent-A-Center, Inc. v. Puritan Farms 2nd, Inc.*, 41 N.Y.S.2d 420, 361 N.E.2d 1015, 393 N.Y.S.2d 365 (N.Y. Ct. App. 1977), ruling that liquidated damages will be enforced so long as they do not appear penal in nature.
4 Although the word "guarantee" or "guaranty" is usually used in this context instead of the word "agree," that term is not quite proper because, in legal terms, "guaranty" usually is used in financial transactions and refers to the performance of one party's obligation by another party if the first party (which is primarily responsible for the obligation) does not perform the obligation as required. "Guarantee" is generally used in the context of consumer warranties or other assurances of quality or performance. Some legal writers prefer guaranty both as a verb and as a noun; however, the modern use of guaranty is as a noun, and the modern use of guarantee is as a verb.

   The real contrast is between guarantee and warranty. "Guarantee" relates to the future, in meaning either (i) the act of giving an undertaking with respect to a contract or performance of a legal act that it will be duly carried out or (ii) something given or existing as security to fulfill a future engagement or condition subsequent. "Warranty" relates to the present or past and is either (a) a covenant connected to a conveyance of real property by which the seller warrants the status of the title conveyed, (b) an assurance given by the seller of goods that the seller will be answerable for their possession of some qualities attributed to them, or (c) in an insurance or other contract, a party's engagement that certain statements are true or that certain conditions will be fulfilled.

5 See *Byrne v. Boadle*, 2 H. & C. 722, 159 Eng. Rep. 299 (1863), holding that barrels do not fall out of warehouses by themselves and an injured passerby should not bear the burden of demonstrating that a warehouseman has behaved negligently if a barrel has dislodged and injured the plaintiff.

6 See *Fletcher v. Trademark Construction, Inc.*, 80 P.3d 725 (Alaska 2003), disallowing a subcontractor's claim for damages as "insufficiently established" and noting that "although a contractor need not prove damages with mathematical precision, it may only recover those damages which it proves with certainty."

Chapter 18

The EPC Contractor's Warranties of the Work

Under the Code of Hammurabi, a builder (and even his family members) could be put to death for construction defects that caused fatalities. Perhaps as a legacy of Hammurabi's perspective on civil works, many states in the United States have statutes that hold architects and builders liable for negligent practices and often have statutes of "repose" or "decennial liability" that "terminate" liability after some predetermined period (often 10 years) from the completion of a project.[1] Pursuing a claim based upon a builder's negligence requires the owner to demonstrate that a builder was negligent. Consequently, it has become common practice for owners to negotiate contractual provisions for warranties of work so that the owner will not have to demonstrate that the EPC contractor's negligence has been involved.

Although an owner will usually require the EPC contractor to design its facility to last for several decades, the EPC contractor will typically limit its responsibility problems to defects that arise and are discovered during the facility's first year or two of commercial operation, although in some industries (such as solar power) warranties of 20 or more years are not uncommon. This warranty from the EPC contractor should cover defects and deficiencies in any part of the work (from computer software to coal crushers, for instance). The warranty should cover equipment, services and omissions. In the case of equipment, a warranty should state that all equipment will be new, of first-rate quality and fit for its intended use. Lawyers often argue over the scope of the phrase "intended use." The EPC contractor's lawyer will often argue that the EPC contractor cannot know what the owner intends for the use of a particular item and, therefore, such a statement cannot be part of the warranty. For instance, an owner could take a spare gauge supplied by the EPC contractor for a natural gas-fired power plant and use the gauge in a nuclear plant even though the gauge might not be designed for such use. There is extensive case law that addresses whether or not manufacturers and vendors of (primarily consumer) products should be responsible for injuries and damage caused by the users of their products (such as the case of a portable electric generator designed for stationary use but installed on a boat, which subsequently causes an explosion because its fuel tank has corroded as a result of being in a marine environment that it was not designed for). In order not to waste negotiation time with endless discussion on this issue, a fair compromise is to agree that all equipment will be fit for use in a facility of a type similar to the facility being built so long as its equipment is operated in accordance with prudent industry practices (see Chapter 6 for a discussion of prudent industry practices).

In the case of services, the warranty should state that all services will be performed properly in accordance with prudent industry standards.

In the case of omissions, the warranty should state that the EPC contractor has supplied all items required or contemplated by the functional specification. If a dispute develops as to whether an item should have been included in the work by the EPC contractor, the owner, as usual, may need to rely upon the wording of the intent clause of the EPC contract if the functional specification is not specific enough to determine whether or not the item should be supplied. For instance, suppose the owner anticipated that remotely located pipeline valves could be operated by electrical signals transmitted by wire as well as by radio control should the radio malfunction. On the other hand, the EPC contractor might have presumed that the owner will dispatch an attendant to adjust the remotely located valves manually if its radio malfunctions. Thus, the EPC contractor did not spend the extra money involved in running electrical wires to these valves. If the functional specification is not specific as to whether failure to include wire control is an omission that will be covered by the warranty or whether a wire connection is beyond the scope of the functional specification (for which an additional price must be paid by the owner) resolution of the issue will turn upon the intent clause of the EPC contract. In a dispute such as this, a judge or arbitrator must attempt to determine whether or not the owner could have reasonably expected the cabling work to be included in the EPC contract price because a facility designed in accordance with prudent practices would have such a "hardwired" cable system in place to serve as a backup system should its radio malfunction.

The owner will be able to invoke the warranty's protection when defects or deficiencies arise. Unfortunately, what constitutes a "defect" or "deficiency" is debatable. To avoid a debate, EPC contractors often prefer that the EPC contract define exactly what the owner means by a "defect" or "deficiency." This, too, can present a dilemma for the owner. Often, the more precise a definition becomes, the higher the probability that something will fall beyond the purview of the definition. Thus, by trying to define "defect" and "deficiency," certain items may be unintentionally excluded. By agreeing to this approach the owner could be put in the position of having no warranty coverage if an item falls outside the scope of the definition. For instance, the following is a typical definition of "defect":

> "**Defect**" means any condition, including a design-life defect, preventing the materials in question from operating in compliance with the contract specifications, except where such condition results from (i) normal wear and tear of such materials, (ii) owner's failure to operate such materials in accordance with the contractor's and manufacturer's operating manual relating thereto, or (iii) further defects resulting from the failure by owner to provide contractor with an opportunity to correct such condition.

The problem with this definition from the owner's point of view is that the functional specification may not even specify how the item or material in question is supposed to operate. In that case, the owner may be required to bear the burden of demonstrating that a "condition" exists and that this condition is preventing proper operation of the facility.[2] It is much simpler to adopt the approach of not attempting to define "defect." The owner then can simply claim that the facility is not operating properly, without being compelled to show that a "defect" exists. To address the EPC contractor's concerns, the EPC contract can make clear that the warranty does not cover normal wear and tear and that actions such as operation of the equipment in contravention of procedures set forth in operating manuals supplied by the EPC contractor or its vendors will invalidate warranty coverage. (The owner should always clarify that if an allowance for "normal wear and tear" is included in the warranty provisions then the EPC contractor's relief from the warranty be limited to "wear and tear that can be expected of properly designed and machined parts made of appropriate alloys of good quality" to make sure that if a part wears because it was not properly designed or fabricated the owner will be covered for this problem under the EPC contractor's warranty.) As a legal matter, the owner should not permit the EPC contractor to structure warranty exclusions as prerequisites or conditions to the effectiveness of warranty coverage but rather as limitations and exclusions to the applicability of the warranty. If a limitation is improperly worded as a condition to a warranty's continued effectiveness, potentially all warranty coverage could permanently lapse and, theoretically, five minutes of operation of a facility outside the operating range prescribed in the EPC contractor's operating manual could permanently invalidate an entire facility's warranty.

Hidden or Latent Defects

Many EPC contracts will provide that the warranty period for defects that cannot be discerned from operations or a visual inspection continues for a period longer than the warranty period for items that are readily apparent or likely to exhibit themselves during the agreed-upon warranty period. Provisions like these are important to include for items that are too difficult to inspect during operations and whose inspection must wait until the machinery that houses them is "opened up" during an overhaul or periodic inspection.

Subcontractor Warranties

Most, if not all, of the equipment and services that the EPC contractor purchases will have warranty coverage from their providers. It is in the EPC contractor's self-interest to ensure that the duration and scope of these warranties is at least commensurate with the duration and scope of its own warranty under the EPC contract; otherwise, since the EPC contractor will be responsible under the EPC contract to remedy any defect or deficiency involving the facility, the EPC contractor will have to cover any gaps in warranty coverage between the warranty provisions of the EPC contract and the warranty provisions of any of its subcontracts. This "flow-down" principle sounds simple but is not always so simple in practice because the EPC contractor will usually not select its subcontractors until after it has entered into the EPC contract because the EPC contractor does not want to have any commitments to subcontractors until it has a commitment from the owner. In addition, the EPC contractor will usually want to create competition among subcontractors before selecting them to make certain the most competitive terms and conditions are obtained.

Superior Vendor Warranties

Occasionally, the EPC contractor (especially in solar power projects) will obtain more favorable warranty terms from some of its equipment vendors than the EPC contractor has given the owner. To address these occurrences, EPC contracts typically contain provisions requiring the EPC contractor to assign any subcontracts containing more favorable warranties than those contained in the EPC contract to the owner at the expiration of the EPC contract's warranty coverage. Other than in this special case, the EPC contractor should be completely responsible for all warranty coverage related to the facility and the EPC contract should make that clear.

"Pre-Ordered" Equipment

In natural gas-fired power plants it is common that the owner will have purchased the combustion turbine and its generator before signing the EPC contract. In a situation like that, it is best (although sometimes more costly for the owner) that the owner come to an arrangement with the EPC contractor that provides that the combustion turbine and its generator will be covered by the EPC contract's warranty even though the owner (and not the EPC contractor) purchased the equipment (as was discussed in Chapter 10). This can be achieved by a simple assignment of the combustion turbine generator purchase contract from the owner to the EPC contractor. This arrangement is prudent because it relieves the owner of responsibility for determining the cause of a defect or deficiency in its facility if the defect or deficiency is not initially apparent. For example, suppose the facility's combustion turbine tends to shut down inexplicably when it is set to produce power at a low output and the owner does not know what is causing the problem. Now, suppose the owner calls the turbine vendor to complain about the problem and the vendor posits that the problem must be a fuel forwarding malfunction and the owner should call the EPC contractor who installed the fuel forwarding pumps. However, what will the owner do if it contacts the EPC contractor about the problem and the EPC contractor informs the owner that the fuel forwarding system is in order and that the combustion turbine is the source of the problem? To avoid this dance of the whip-poor-wills, the owner should simply make the EPC contractor responsible for all facility problems.

"Serial" Defects

Some facilities (such as solar or wind power plants or water desalination facilities) are essentially massive replications of a single set of components. Solar panels, wind turbine blades and specialty piping are all components that may be present by the hundreds or thousands depending upon the size of the facility in question. Often owners are able to negotiate so-called "serial" defect warranties to provide that if a large enough percentage of a particular component proves defective, the owner may be able to extend the warranty period for all those particular components at no cost or sometimes may even negotiate a provision requiring replacement by the EPC contractor of all remaining similar components at the facility even if they have not exhibited any defect.

Locating Defects and Deficiencies

The EPC contract should make clear that the EPC contractor has the duty, at its own expense, to search for and locate the "root cause" of any problem. This responsibility should also include so-called "in and out" costs. These are expenses associated with disassembling and reassembling equipment that may or may not be associated with the warranty defect itself (for instance, a pipe rack may have to be disassembled to permit access to a burst boiler tube). The EPC contractor should also be made responsible for packaging, shipping and customs costs for items that cannot be repaired on site.

Remedies for Breach of General Warranty

Once it is determined that a problem exists, the EPC contractor should be obligated to make repairs as quickly as possible, particularly if the owner cannot operate its facility until the repair is made. Usually, after a facility has been out of service a certain number of days (typically 45), the owner will be able to begin collecting business interruption insurance (see Chapter 23) but this insurance usually only covers the owner's costs and not the owner's lost profits (profits are usually prohibitively expensive to insure). So, making a claim under this insurance may not put the owner back in the position the owner would have been in had the problem been rectified promptly.

The EPC contract should impose a deadline on the EPC contractor for effecting reparations. Giving the EPC contractor a "reasonable" period in which to remedy problems is not advisable. The EPC contractor and the owner may have differing understandings of what is a "reasonable" period. If the EPC contractor does not complete reparations within an allotted time specified in the EPC contract, the owner should be able to take matters into its own hands and fix the item itself (or use another contractor) and charge the EPC contractor for costs associated with this remedial action. Of course, not all defects may be susceptible to cure in the period prescribed by the EPC contract. For instance, if the EPC contract provides that a cure must be completed within seven days after the EPC contractor has been notified of the problem by the owner, the EPC contractor will obviously not be able to comply within this timeframe if a long lead time item like a transformer must be replaced (which could easily take 6 to 12 months to fabricate and deliver). To handle such cases, the EPC contract usually will provide that if correction of the problem is not commercially feasible during the remedial period enumerated in the EPC contract, the period will be extended for as long as reasonably necessary to allow the EPC contractor to implement the proposed solution so long as the EPC contractor pursues the remedy diligently. Thus, as long as the EPC contractor continues to act diligently in pursuit of the plan of action, the remedial deadline will be tolled accordingly. Additionally, the owner and the EPC contractor may have different perspectives of what is "a diligent pursuit of a remedy." For instance, consider a faulty transformer in a power plant. The owner may believe the EPC contractor should rent a transformer for the owner until a new transformer can be installed. The EPC contractor may believe that the owner's facility should merely run at a reduced output if the malfunctioning transformer is only one of several in the power plant. The owner may also believe that the EPC contractor should locate a vendor that can offer expedited service in fabrication of the new transformer. The EPC contractor may want to use the original transformer vendor, who will supply a new transformer free of charge under the vendor's warranty to the EPC contractor but perhaps on the same schedule as set forth in the original purchase order between the EPC contractor and the transformer vendor.

In all of the above cases, because there can be differences of opinion and differing financial incentives (the owner losing profits and the EPC contractor distracting its personnel from other endeavors), the owner should require the EPC contractor to submit promptly a reasonable warranty repair plan for the owner's approval when any warranty work cannot be carried out within the time dictated by the EPC contract. This should help align expectations of the parties while warranty work is being carried out.

Scheduling of Warranty Work

Often, while a problem may not prevent safe operation, a facility may have to be taken out of service or run at a reduced level for its reparation. The owner should have the power to decide when the EPC contractor can perform warranty work (given the owner's obligations to its offtaker or user or its reduced profit potential if it has no customers and is selling output into the market). Therefore, the owner should retain flexibility to relegate warranty work to non-peak times such as nights and weekends (or "shoulder months" such as October and April when electricity demand is traditionally low in most of the United States in the case of a power plant) or even upcoming scheduled or unscheduled maintenance outages of the facility. Most EPC contractors will not object to a statement in the contract that they agree to work in good faith with the owner in the scheduling of warranty work. Since the owner may be liable to pay liquidated damages to its offtaker for service interruptions, sometimes the owner will require the EPC contractor to pay daily liquidated damages to the owner if the EPC contractor fails to remedy a problem within the period prescribed by the EPC contract. While most EPC contractors will vehemently resist this, they usually will agree to pay daily liquidated damages (typically a nominal amount only) if they fail to commence warranty work, or to investigate the problem diligently, or to perform as required by provisions of the EPC contract. Ordinarily, these damages will cease to accrue once the EPC contractor has begun to take appropriate action to commence reparations.

Occasionally, EPC contracts will provide that the owner or EPC contractor can elect to forgo a warranty repair and the EPC contractor will instead be obligated to pay to the owner the diminution in value of the facility as a consequence of the problem that has arisen. A provision like this should be avoided because it has the potential to create disputes over the diminution in value of the facility and give the EPC contractor a defense against performing very complicated or intricate warranty work. The parties are always free to agree upon any solution for a warranty problem at the time it arises irrespective of what the EPC contract says. Therefore, a provision like this is not necessary, or if one is to be included in the EPC contract, it is best if it gives only the owner (and not the EPC contractor) this "buy out" right.

Warranty Exclusions

The Industrial Revolution gave birth to sophisticated machinery and consumer products that had the potential to maim and injure workers and consumers as never before. Courts and legislatures struggled for decades with the issue of how to compensate injured parties. From birth defects to asbestosis, products of the twentieth century could affect not only individuals but entire segments of society. In the United States, a vast body of law has developed to give these victims some redress from the parties responsible for the infliction of their harm. Doctrines of law began to hold manufacturers and sellers responsible on the basis of "implicit warranties" of fitness for the use intended or even "strict liability", that is, liability to which there simply is no defense available once the plaintiff is able to show that the defendant's product has caused its injuries. In an attempt to avoid being subject to implied warranties deemed to have been made to a user irrespective of whether or not they were expressly contained in a contract of sale, manufacturers began to include language in their contracts of sale that stated that implied warranties were expressly excluded from the terms and were not applicable. As a matter of public policy, especially when there is perceived to be unequal bargaining power between the parties (as in consumer transactions), such exclusions and denials are generally ineffective in most jurisdictions in the United States.

As a general practice, EPC contractors insist on exculpatory and exclusionary provisions that attempt to negate implied warranties, although implied warranties actually developed to cover consumer transactions and not large commercial transactions. Even so, provisions that attempt to waive rights existing under common law are often of questionable enforceability in most jurisdictions in the United States (although the parties to an EPC contract generally do have equal bargaining power, which may increase the chances of successful enforcement of such waiver provisions in a court). Finally, if the warranty clause states that the EPC contractor will remedy defects, the owner is probably protected without having to resort to implied warranties of fitness for use. The owner is probably protected because the EPC contractor will remedy the problem in question and any harm caused by the problem (such as injuries to workers or third parties) will be covered by workmen's compensation or third-party liability insurance rather than by the EPC contractor (see Chapter 23). Furthermore, damage to property, such as so-called "collateral" damage that occurs when another part of the owner's facility is damaged as a result of a problem (such as a smoke stack collapsing on a facility's fuel tanks) will be covered by the owner's property insurance. As will be discussed in Chapter 20, the EPC contractor will almost always disclaim liability for any consequential damages, such as lost profits, that the owner suffers as a result of any problem. Therefore, honoring the EPC contractor's request to include a waiver of the EPC contractor's liability for implied warranties of fitness for use is probably acceptable because the warranty in the EPC contract will be crafted to try to cover all circumstances about which the owner should be concerned.

The EPC contractor will also usually exclude certain items from its warranty coverage. Normal wear and tear of equipment and depletion of items commonly referred to as "consumables" (fluids, filters and the like) will be excluded from the scope of the EPC contractor's warranty. The EPC contractor will also exclude deleterious conditions created by the owner's improper operation of its facility, including the owner's use of inappropriate inputs such as contaminated fuel or feedstocks, or the use of spare parts that have not been stored in accordance with the EPC contractor's instructions.

Limitation of Warranties

In addition to exclusions from the warranty coverage, the EPC contract will limit the duration of warranty coverage. Generally, the warranty will survive for one or two years from commencement of commercial operation (usually substantial completion or provisional acceptance) of the facility. In fact, at what point the warranty coverage is to commence can become an issue. Often the EPC contractor will request that warranty coverage takes effect when the facility is ready for performance testing in case the owner delays testing for some reason. This is a reasonable request so long as this provision also provides that the warranty period will be extended if the performance tests demonstrate that the facility is not ready for service once it actually does undergo performance testing.

The owner must be especially vigilant if it has purchased equipment directly from a manufacturer rather than the EPC contractor, as was discussed in Chapter 10. Manufacturers usually start warranty clocks running from the time their equipment is delivered (or shipped). Thus, in the case of a combustion turbine purchased by the owner from the vendor that is covered by an 18-month warranty but where defects are not included as part of the EPC contractor's warranty coverage under the EPC contract, the owner could be placed in a position in which the owner has little or no warranty coverage on the turbine if the manufacturer delivers the combustion turbine to the site on time but the EPC contractor takes 17 or 18 months to install it because other (unrelated) problems arise. Again, this problem can be avoided by including the combustion turbine in the EPC contractor's warranty.

EPC contractors generally do not want to give "evergreen" warranties that will survive indefinitely. In other words, if the EPC contractor replaces or fixes an item, it will usually warrant the item's performance for an additional period equal to the original warranty period for that item. However, in cases of a chronic failure, the EPC contractor will usually not obligate itself under the EPC contract to replace or fix the item indefinitely. Typically, the EPC contract contains some ultimate cut-off date after which the EPC contractor will have no further liability to the owner for replacement or repairs of an item that has repeated problems. However, as a practical matter, most responsible EPC contractors and vendors will continue to repair or replace an item whether or not they are obligated to do so under the EPC contract if the item fails to perform as required. One exception to this "sunset" warranty concept is usually made for new technology, which an EPC contractor or vendor often must support with an "evergreen" warranty (common in solar plants) in order to promote the technology's initial market penetration. Other exceptions that can extend the warranty period are so-called "serial" defects provisions, which provide that in case some threshold number or percentage of a common component (such as a solar cell of which a facility may have hundreds or thousands installed) fail, the warranty for that item becomes evergreen for all such like items.

Warranty Bonds

In some cases, the owner will ask the EPC contractor to post a bond or letter of credit during the warranty period to secure the EPC contractor's obligation to perform warranty work. A warranty bond from a bonding company is intended to cover the EPC contractor's "faithful" performance of all required warranty work. On the other hand, a letter of credit from a bank allows the owner to draw up to a fixed amount of cash (typically 5 percent of the EPC contract price) if the EPC contractor has failed to perform any warranty work. Either of these practices is helpful because the EPC contractor will usually have been fully paid before the warranty period commences and may have little economic incentive to execute warranty work aggressively, especially if the EPC contractor is not doing other work for the owner on other projects or does not agree with the owner about the presence of a problem. Suppose a power plant passes all its performance and emission tests, but the EPC contractor has used pressure-treated wood in constructing the power plant's cooling towers. After six months, arsenic used from treating the wood begins to leach into the cooling tower water and the plant then violates its water discharge permits. Suppose further that the functional specification did not expressly state that the EPC contractor was not supposed to use treated wood in the construction of the cooling tower. Although there may be many ways to solve this problem, the EPC contractor may be much more willing to participate in resolution of the problem if it knows that the owner can draw upon a letter of credit at any time (rather than sue the EPC contractor or the bonding company or both of them) and then the EPC contractor will have to fight with the owner to "reclaim" the EPC contractor's cash if the EPC contractor believes the cost of the owner's self-implemented solution was unreasonably expensive. (See Chapter 19 for a discussion of letters of credit.)

Notes

1 See *Kozikowski v. Toll Bros., Inc.*, 354 F.3d 16 (1st Cir. 2003), upholding a statute of repose to bar an action.
2 As a general matter, in power plants the hot gas or steam path in a turbine is precisely engineered with minute tolerances and therefore this equipment must be safeguarded because it is often the most expensive and difficult part of the power plant to repair (sometimes other pieces of equipment will even be sacrificed in order to prevent damage to a turbine). For example, steam turbines are designed for pure steam to run through them. If a "condition" arises so that some steam condenses into a water droplet before it enters the steam turbine, the droplet would have the potential to "nick" or chip the steam turbine's blades and lead to problems that could entail opening the entire steam turbine for repair (which may not be good for the owner because many combined cycle power plants cannot operate with the steam turbine opened up because the steam created by the heat recovery steam generator cannot be condensed into water in any other portion of the plant). Thus, if there is any indication that the steam about to flow though the steam turbine is not hot enough, the operator or the plant control system may immediately turn the steam turbine off to prevent damage to it. However, this unexpected disruption of service can have many other (potentially expensive) consequences (such as boiler tubes cooling more quickly than their rated strength, leading to their rupture and cracking). So, the damage to the turbine block should be easy for the owner to demonstrate but the existence of the condensation condition as a "defect" may be difficult to prove, which is why a definition of "defect" is not desirable.

Chapter 19

Security for the EPC Contractor's Performance

Performance Bonds

From shipbuilding to highway construction, performance bonding is a common practice. It involves a third party either posting cash or acting as a surety to ensure that a contracting party will perform its obligations. If the obligor does not perform, the obligee can draw the cash or demand that the surety perform if a bond has been executed. Like the law of admiralty and realty, the law of suretyship in the United States is well established and it would seem that in the case of an EPC contract, the owner would simply need to inform the surety that the EPC contractor has failed to perform and the surety will dutifully assume the EPC contractor's performance obligations. Unfortunately, as owners sometimes find out, this is not always the case.

If the surety is still in business and has the resources to perform the EPC contract at the time of the owner's demand (which, in the case of large projects, can be years after the EPC contract has been signed), the surety may still may have many defenses to its performance that are "personal" to the surety.[1] Many common law protections have developed for sureties and guarantors. For instance, variation of the underlying contract between the obligor and obligee is a defense to the surety's performance obligation if the variations have been made without the consent of the surety.[2] The theory, here, logically, is that the scope of the surety's obligations cannot be expanded without the surety's consent. Thus, change orders that enlarge the scope of the functional specification or raise the required level of performance of the facility have the potential to release the surety from its performance obligations. Furthermore, a surety bond as written often requires that the owner exhaust all its remedies against the obligor before attempting to seek performance from the surety.

While performance bonds may actually work quite well on modestly sized civil works projects, given the complexity and cost of large facilities like power, LNG and processing plants, performance bonds from sureties are seldom used (except in cases of contracting with government agencies and in these instances performance bonds and their format are often required by regulation). Obviously, for an integrated, international project of large proportions whose cost may be billions of dollars, a bond from a local surety may not be appropriate.

Given that owners and financiers want more than just the right to lodge a separate lawsuit against a surety if an EPC contractor fails to perform its obligations, other alternatives in lieu of performance bonds have developed for large projects. The efficacy of these alternatives, discussed below, varies.

Parent Guarantors

As was discussed in Chapter 3, since EPC contractors usually set up subsidiaries to carry out projects, it is typical for an owner to require a guaranty of the EPC contractor's performance by one of the EPC contractor's parent entities. This entity will typically be the EPC contractor's ultimate parent entity (unless one of the intermediate holding companies in the chain of ownership of the EPC contractor has the financial and technical resources to perform the responsibilities of the EPC contractor under the EPC contract).

EPC contractors often request a certain period after the EPC contract has been signed to deliver their parent's signed guaranty. This should be acceptable to the owner as long as the period is relatively short and the EPC contractor agrees that the owner will have no obligations under the EPC contract unless the guaranty is delivered within the allotted timeframe and the owner can terminate the EPC contract if it is not received within the allotted time.

owners should be cognizant that the "parent" from which they will be requesting a parent guaranty may have its own officers and legal counsel who may be involved in negotiation of the parent guaranty. It is important that an owner insist that the form of the proposed parent guaranty be delivered by the EPC contractor promptly to its appropriate parent personnel for review. These personnel may not be enthusiastic about spending time reviewing this document until they are convinced that their subsidiary is likely to enter into the EPC contract. These officers may also have concerns and policies regarding the parent's credit support of its subsidiary. Thus, the practice of promptly forwarding the parent guaranty to the parent can help avoid consequences that can have a serious (or even fatal) impact on the owner's project. For instance, many parent entities have policies against giving guaranties unless the guaranties have an expiration date and an overall limitation of liability. Both of these practices will generally be unacceptable to the owners' lenders if the owner is financing its project because projects can run into delays and overruns and lenders will be looking to the EPC contractor (and its parent) to honor the EPC contractor's obligations under the EPC contract no matter what circumstances arise. Unfortunately, in this situation, the only compromise may be for the owner itself to assume responsibility for the EPC contractor's performance if these limits are reached because the EPC contractor's parent limits its liability. Obviously, such a backstop by the owner could take a significant toll on the owner's rate of return and overall credit standing because the owner may be requested at some future point to invest more equity in its project than it expected if the EPC contractor does not perform and its parent exhausts its liability exposure to the owner.

As was explained in Chapter 3, it must be understood, perhaps contrary to common understanding, that there is no such thing as a company operating under an "umbrella" or "group" of companies. With very few exceptions (such as fraud), properly incorporated and managed legal entities are not responsible for the debts and obligations of other properly incorporated and managed legal entities, even if they own them (directly or indirectly through a chain of other entities), unless they have agreed by contract to assume these liabilities. Perhaps it is best to think of the analogous situation that parents are not responsible for the debts of their adult children. It is easy for an uninformed owner to fall into the "consolidated financial statement" trap. When an EPC contractor sends the owner the consolidated financial statements of its parent entity, this means absolutely nothing. Each entity controlled by the EPC contractor's ultimate parent entity may have its own assets and liabilities and these assets and liabilities may or may not be beyond the reach of their parent entity, as was discussed in Chapter 13, whether or not their accounting practice is to show consolidated financial statements.

Other Security

A further option for the beneficiary of a parent guaranty would be to request that the guarantor actually pledge some collateral to secure its obligations to the owner under the guaranty. Thus, the owner might ask the EPC contractor's parent to pledge equipment or other assets (such as its ownership interest in the EPC contractor) that the owner could foreclose upon if the EPC contractor's parent does not honor its guaranty. With collateral, the owner will probably be in a better position against the EPC contractor's parent.

Collection vs. Payment

In general, a beneficiary of a guaranty will be seeking payment of a sum of money from a guarantor. Sometimes the terms of the guaranty will require that the beneficiary obtain a judgment or arbitral award against the debtor (EPC contractor in this case) and that the beneficiary (owner) may only proceed against the guarantor if the beneficiary (owner) is unable to collect the judgment or award from the debtor (EPC contractor). This is often known as a "guaranty of collection." In other cases, a guarantor may agree to make payment or even perform an obligation if the obligor fails to do so. This is known as a "guaranty of payment."

Forum

In any of the above cases, once the underlying facts are settled (generally in a proceeding between the debtor and the beneficiary) enforcing guaranties in U.S. courts is relatively simple, and sometimes even an abbreviated procedure is available. Therefore, beneficiaries usually prefer to proceed against guarantors in court because the law concerning guaranties is well established. Arbitration, in which there is always the risk of an unexpected result, is typically avoided by beneficiaries of guaranties. As noted in Chapter 3, issues can arise if the EPC contractor calls for arbitration and the parent guarantor calls for court proceedings. If the owner is comfortable with this dichotomy, the owner should ensure that the parent guaranty provides that the parent guarantor will be bound by any underlying arbitration between the owner and the EPC contractor regarding the EPC contract in order to avoid "re-litigating" a controversy with the parent guarantor.[3]

Letters of Credit

"Direct Pay"

Letters of credit developed as a "third party" mechanism to ensure that an obligor's payment would be made once performance had been rendered. The performing party could simply show a document such as a bill of lading or a warehouseman's receipt to the issuer of the letter of credit and the issuer of the letter of credit would then immediately make payment to the performing party upon the issuer's "sight" of the specified documentary evidence. These "documentary" or "sight" letters of credit allowed counterparties to transact business with each other with minimum payment risk so long as the seller was comfortable with the creditworthiness of the issuer of the letter of credit, typically a bank. Thus, a simple credit investigation into the bank's finances (such as an inquiry made to a credit rating agency like Dun & Bradstreet or Standard & Poor's or Moody's) could tell a potential seller of goods what the risk was that the issuing bank would not have funds to make payment to the bearer of the letter of credit upon presentation of the letter (and relevant documents) for payment.

Essentially, unless the necessary documents were not in order upon their presentation by the beneficiary (seller of the goods) to the issuer of the letter of credit, the issuer would pay "directly" under the letter of credit. In fact, there are almost no defenses to a letter of credit's payment if the specified documents are in order.

The ICC has developed a code (the Uniform Customs and Practice for Documentary Credits) to govern the administration and payment of letters of credit. Most banks will make their letters of credit subject to these rules. Letters of credit have probably had the effect of helping to promote international trade because sellers have been able to rely on the credit of financial institutions with whom they are familiar when sellers seek payment for the goods they have sold. By utilizing a letter of credit, a seller does not have to sell goods on "open account" to a buyer in a foreign jurisdiction and worry that it will be difficult and expensive (if not impossible) to "track down" the buyer if the buyer fails to make payment. The burden of collecting from the buyer will be borne by the bank that issued the letter of credit for the account of the buyer if the buyer does not repay the bank that has advanced the money to the seller on the buyer's behalf. Naturally, the bank will charge an account party a fee to issue a letter of credit and also charge interest on the amount of the letter of credit from the time the bank makes payment to the beneficiary of the letter of credit until the time the account party (buyer) reimburses the bank in full for the bank's payment to the beneficiary of the letter of credit. Banks usually also charge a "commitment fee" on the amount of the letter of credit from the time the letter of credit is issued until the time the letter of credit is drawn by its beneficiary.

In international letter of credit transactions, frequently more than one bank is involved. A foreign buyer may have a relationship with a local bank that is familiar with the buyer and its creditworthiness. This local bank may be willing to issue a letter of credit on behalf of the buyer. In fact, even if the buyer does not have stellar credit, the local bank may be willing to post a letter of credit for the buyer if the buyer posts cash or other collateral in favor of the bank in the amount of the credit letter in order to secure the buyer's obligation to reimburse the bank once the letter is drawn upon by the beneficiary. However, the seller might not be familiar with the local bank and might be hesitant about accepting this local bank's letter of credit, especially if the bank is not doing business in the seller's home jurisdiction. For this reason, banks doing business internationally have developed relationships among themselves. The buyer's local bank may have a relationship with a local bank in the seller's home territory. In this case, the buyer's bank may issue a letter of credit to the seller's local bank which will, in turn, issue another letter of credit to the seller (or a process of "confirmation" may also be used whereby a bank in the seller's home territory will "confirm" to the seller that the foreign bank's letter of credit will be paid).

Eligible Banks

As will be seen below, if letters of credit are to be used in connection with an EPC contract, the concept of an "eligible issuing bank" should be built into the EPC contract. An "eligible issuing bank" should be defined as a bank that has and maintains the credit rating and minimum balance of capital that the owner believes are acceptable from a credit-risk point of view. For instance, a local home savings bank might have a good credit rating but not the liquidity or reserves required to support a large letter of credit. Trade finance is a highly specialized business line for banks, so account parties and beneficiaries should make sure they are dealing with a bank that has experience in this type of business. An "eligible issuing bank" should also have a branch in one of the world's major banking centers like New York, London or Hong Kong. A beneficiary should never accept a letter of credit unless it is drawable at the counters of a bank in a major banking city in a "hard" currency such as dollars or euros. Obviously, it is acceptable if the letter of credit can be drawn at "foreign" counters in addition to a counter in a major banking center, but a beneficiary should not have to travel to a country that has become politically, militarily or financially "unstable" to try to collect its money in a local currency.

Standby Letters of Credit

"Standby" letters of credit have developed to address situations in which a buyer is expected to make payment itself (rather than the letter of credit being drawn upon by the seller) but fails to do so. Standby letters of credit are typically less costly than "direct pay" letters of credit (because banks do not expect to use their own funds to pay under them) and are typically obtained in situations in which the counterparties to an underlying contract already have an established relationship with, or are fairly comfortable with, each other's creditworthiness, such as sales of natural gas on pipelines.

When are letters of credit used under an EPC contract? Typically, to "support" the EPC contractor's obligation to pay liquidated damages to the owner for delays in completion of the facility or shortfalls in the expected performance of the facility. From an owner's point of view, it is easier to present a demand for payment of liquidated damages to a bank under a letter of credit than to demand payment of liquidated damages from a disgruntled EPC contractor or its parent entity (or a surety). A letter of credit will give the owner the option to satisfy the EPC contractor's obligations to pay liquidated damages quickly and easily should the EPC contractor, or its parent or surety, not make a payment of liquidated damages when they are due or in dispute.

Sometimes letters of credit are used to assure more than just payment of liquidated damages but also warranty or other obligations under the EPC contract. The posting of a letter of credit in favor of the owner may raise the EPC contract price because the bank fees associated with a letter of credit will increase the EPC contractor's cost of performance, and an outstanding letter of credit could restrict the EPC contractor's liquidity for future borrowing or its capacity to have sureties post bonds on its behalf. In fact, letters of credit are not as typical in projects in the United States as they are elsewhere. In a typical EPC contract structured with a letter of credit to "backstop" the EPC contractor's obligation to pay liquidated damages, the amount of the letter might be equal to 10 percent of the EPC contract price, while liquidated damages payable under the EPC contract might be capped at 25 percent of the EPC contract price (as was discussed in Chapter 17 and will be discussed again in Chapter 20). In this case the owner will still be "at risk" for more than half of the potential amount of liquidated damages. Even if a larger letter is issued, it is unlikely that the letter of credit will ever approach the full EPC contract price so it is imperative for the owner to have another form of performance security that will cover the entire value of the work—such as a guaranty from the EPC contractor's ultimate parent entity or a surety bond.

Furthermore, since there is likely to be a lengthy period between signing of the EPC contract when the letter of credit is posted and drawing under the letter (if necessary) perhaps years later when the project is completed, it is important that the owner make the insolvency (or even a significant "downgrade" in the credit rating) of the bank that issued the letter of credit an event of default under the EPC contract. This will give the owner leverage to require the EPC contractor to obtain another letter of credit in lieu of the letter issued by the bank in financial trouble.

Occasionally, EPC contractors will suggest that they not be required to post a letter of credit until a much later date in the project or unless the EPC contractor misses one of the "guaranteed" dates such as the substantial completion date. They will base this argument on the fact that this is a more cost-effective solution for the owner, who does not really need the letter of credit until later in the project or until a "guaranteed" date is missed (which is true), and the EPC contractor should not needlessly be paying commitment fees to keep a letter of credit outstanding that cannot be drawn upon by the owner until a "guaranteed" date is missed. Thus, the EPC contractor should not be required to diminish its corporate borrowing or bonding capacity by posting a letter of credit that is not even drawable for most of the time it will be outstanding. The owner should not accept this proposal. If a project is not going smoothly and the EPC contractor is already in breach of the EPC contract for other reasons, the EPC contractor is likely not to be concerned about its future breach of this provision requiring the posting of a letter of credit if the EPC contractor does miss a "guaranteed" date.

Although it is true the owner can stop making payments on the EPC contract price to the EPC contractor if the EPC contractor does not satisfy its obligation to have a bank issue a letter of credit in favor of the owner when the EPC contract requires, if the owner takes this course of action, the EPC contractor may have no funds to pay its subcontractors and work will probably stop and create further problems. Also, if the EPC contractor is having financial difficulties at the time it is required to post a letter of credit under the EPC contract, the EPC contractor may not be able to find a bank willing to extend credit to the EPC contractor and post the letter of credit on the EPC contractor's behalf. Whatever the case, if the owner (or its lenders) have decided that it (or they) want a letter of credit in connection with the EPC contract, the EPC contractor should be required to deliver the letter of credit at, or before, notice to proceed has been given by the owner to the EPC contractor and not later.

Expiration

The owner must be careful to negotiate the proper term of the letter of credit. A letter of credit should not expire before substantial completion of the facility has actually occurred plus some additional period (usually 30 to 60 days) for the owner to draw upon the letter if the EPC contractor does not pay any delay or performance liquidated damages it owes. If a cure period is granted to the EPC contractor and liquidated damages are not payable until the end of the period, then the additional period to draw on the credit letter should be added to the cure period.

One complication that often arises is that most banks have their own policies that prevent them from issuing letters of credit for long durations (usually in excess of 364 days so it is not booked as a long-term liability on its balance sheet). Even if a bank will issue a letter of credit that is valid for more than one year, banks generally require their letters of credit to expire on a predetermined calendar date and not an uncertain date in the future that may never even occur—such as the date of substantial completion of the facility. To avoid issuing a potentially perpetual letter of credit, because the bank generally will insist on a calendar expiration date for its letter of credit, the owner should be sure to include a provision in the letter of credit to the effect that, if the letter is scheduled to expire within 30 days and by that expiration date the completion of the project has not occurred, the owner, unless the bank has extended the expiration date or the EPC contractor has furnished another letter of credit meeting requirements of the EPC contract, can draw upon the expiring letter of credit even if liquidated damages are not payable at that time and hold the funds in escrow until completion occurs.

Bank Guaranties

In the United States, banks cannot guaranty the performance of obligations of another person (unless that person is affiliated with the bank). In other jurisdictions, however, banks can issue guaranties. Sometimes an EPC contractor will suggest furnishing the owner with an "on demand" bank guaranty instead of a letter of credit because this is less costly for the EPC contractor. Banks generally charge less to issue and maintain a bank guaranty than they do to issue a letter of credit.

Usually, the EPC contractor will argue that there is no difference between an on-demand bank guaranty and a letter of credit. Owners should understand that a bank guaranty is not the same as the letter of credit. Guarantors (including bank guarantors) may have defenses to guaranties (which the guarantor will typically be asked to waive in the guaranty agreement, but these waivers can be of questionable enforceability). In juxtaposition, if proper documents are presented by the beneficiary of a letter of credit, there really are no defenses to an issuer's paying a letter of credit (except in extreme cases such as fraud or shipping goods under the flag of a carrier chartered in an embargoed country, for instance). A letter of credit is the next best thing to the owner holding the EPC contractor's cash and, in fact, in some cases (such as the bankruptcy of the EPC contractor) a letter of credit is better than holding cash (because in the case of the EPC contractor's bankruptcy, the court has the power to take the escrowed cash from the owner because the cash is still the EPC contractor's cash even though it is being held by the owner). On the other hand, the letter of credit is between the owner and the EPC contractor's bank and has nothing to do with the EPC contractor and therefore the owner can draw (and usually keep the cash proceeds of) the letter of credit even if the EPC contractor has filed for protection under the U.S. Federal Bankruptcy Code. (See Chapter 13 for a discussion of U.S. bankruptcy principles.) A bank guaranty, a parent guaranty or a surety bond can have the consequence of putting a lawsuit between the owner and the EPC contractor's cash.

Notes

1 See *Federal Ins. Co. v. Broadmoor, LLC,* 2003 WL 282324 (E.D. La. Feb. 10, 2003), ruling that a surety was not bound by a contractor's agreement to arbitrate.
2 See *United States Fid. and Guar. Co. v. Braspetro Oil Serv. Co.,* 219 F. Supp. 2d 403 (S.D.N.Y. 2002), in which sureties refused to perform an off-shore construction project on the basis that the owner had varied the contract work.
3 See *Dunn Indus. Group, Inc. v. City of Sugar Creek,* 112 S.W.3d 421 (Mo. 2003), in which it was held that a guarantor was not bound to submit a dispute to arbitration because the underlying design-build contract contained an arbitration provision. Also see *Intergen N.V. v. Eric F. Grina, Alstom (Switzerland) Limited, and Alstom Power N.V.,* 344 F.3d 134 (1st Cir. 2003), holding that a sponsor was not bound by its SPV's agreement to arbitrate. But see *Choctaw Generation Limited Partnership v. American Home Assurance Company,* 271 F.3d 403 (2d Cir. 2001), in which it was held that an owner's action against a surety that the owner had filed in a court had to be consolidated into an arbitration between the surety and the EPC contractor.

Chapter 20

Limitations on Overall Liability and Contract Expiration

Grave circumstances have the potential to cost the EPC contractor far more money than its expected profit. EPC contractors often attempt to limit their overall liability to owners and not just their liability for liquidated damages for delays in completion and poor performance of facilities (as was discussed in Chapter 17).

Liability can arise in several ways. It can arise by contract (the case in which two parties enter into an agreement between themselves), by operation of law (for example, dumping toxic waste in a landfill is prohibited by statute), or under common law for so-called "tortious" conduct for breaching a duty of care owed to another party (for example, oil leaching from one party's property to another party's property can give rise in the United States to a common law right of indemnity so that the injured party can recover damages). The EPC contractor will propose to limit all these types of liability, not just the EPC contractor's liability arising under the EPC contract (although the "economic loss" doctrine of common law will usually prevent a party to a contract from recovering its economic losses from the other party in tort).[1] It is typical for an owner to agree to this request from the EPC contractor, but often the owner may insist upon several exceptions to this limitation. One notable exception is the case in which the EPC contractor's liability to the owner is covered by insurance and thus the EPC contractor is not "out of pocket" for costs paid by an insurer and therefore its liability should not be reduced by the amount of the insurance proceeds (see Chapter 23). Another case in which the liability limitation will often be disregarded is fines levied against the owner on account of the EPC contractor's actions.

Negligence and Gross Negligence

Yet another instance in most U.S. jurisdictions in which the liability limitation will usually be disregarded by courts is the situation in which the EPC contractor has acted with "gross" negligence. Unlike "negligence," where the EPC contractor has failed to act prudently and breached a legal duty of care not to expose a party to risk of harm, "gross" negligence is often described by U.S. courts as the case in which a party has acted with a "reckless disregard for the consequences of its behavior."[2] Since the burden of proof for the owner to bear in proving gross negligence is more difficult than the burden of proof for negligence, the EPC contractor will generally agree to including this provision in the EPC contract. As a legal matter in most U.S. jurisdictions, even if the EPC contractor will not agree on an exception to the liability limitation for gross negligence, it may be that the owner will still be able to recover damages from the EPC contractor in excess of the liability limitation if the EPC contractor has acted recklessly or in bad faith. Statutes and courts generally will not permit a party (such as the EPC contractor) to absolve itself of liability pursuant to a contractual provision and then act maliciously or recklessly because the party has a contractual provision releasing it from liability for its actions.[3] As noted in Chapter 13, when an EPC contractor is nonchalant or unruly and fails to take reasonable measures to meet the completion date for the work that it has promised in the EPC contract, the EPC contractor may not be able to take refuge in the EPC contract's limitation on the EPC contractor's payment of liquidated damages and limitation on the EPC contractor's overall liability because the EPC contractor's actions may be inconsistent with the intent clause of the EPC contract (if the intent clause has been properly drafted [as was discussed in Chapter 6]). Thus, the owner may be able to invalidate the liability limitation based on the EPC contractor's impertinent behavior.

Types of Damages

"General" or "direct" damages compensate an injured party for its so-called "direct" losses, such as the costs of repairing a frozen valve. "Special" or "indirect" damages compensate a party for losses that arise from the injured party's partial circumstance, such as profits forgone because its facility could not operate while the valve was frozen. The seminal 1854 case of *Hadley v. Baxendale*[1] propounded the tests for "direct" loss (those that arise directly and naturally from breach of contract) and "indirect" loss (those that arise from special circumstances of the parties that were known by them at the time that they entered into the contract). Thus, the terms "direct loss" and "indirect loss" are used by the judiciary of the U.K. to denote categories of losses, but what losses fall within what category depends upon the particular circumstances of the parties to the suit (see "Waiver of Damages" below), and therefore there is no precise definition of each of these categories but rather a conceptional notion of how to categorize a loss. "Direct" loss is often delineated as the natural consequences of a breach of contract that anyone could expect to arise. Under U.K. law, "consequential loss" has no significance; it is not a legal term. Items such as lost profits can be either "direct" or "indirect" loss depending upon the case at hand. Consequently, excluding "indirect" losses from a party's liability to the other party may not always exclude lost profits so the attorneys, without knowledge of the particular situation, drafting a contract must make clear whether lost profit is intended to be excluded from the parties' liability to each other.

Waiver of Damages

Typically, the parties to an EPC contract will waive claims against each other for special damages because these damages have the potential to be huge. There are other types of damages, such as exemplary and punitive damages, in which a U.S. court may award damages to make an example of a defendant or punish its atrocious behavior. Damages of this type are also often waived by parties to EPC contracts, although, as discussed above, these waivers may be against public policy and unenforceable in many U.S. jurisdictions.

Who Owes the Damages and to Whom Are They Owed?

Since both the EPC contractor and the owner are likely to have many officers, directors, affiliates and subsidiaries, it is often prudent to state that the EPC contractor and the owner will look solely to one another in cases in which liability arises. They might further agree that they will prevent any of their affiliates from becoming involved in disputes between them that could have the practical effect of circumventing all the damage limitations and waivers in the EPC contract because these affiliates are not contractually bound by any waivers set forth in the EPC contract since they have not signed the EPC contract. As was noted in Chapter 19, there is no such thing as companies "operating as a group" or companies operating "under the same umbrella." Therefore, legally, a party cannot bind its affiliates unless they themselves intended to be bound somehow, such as by acknowledging the waiver contained in the EPC contract in writing or giving a party to the EPC contract a power of attorney that enables the party to bind the affiliate.

It is also important for the EPC contractor to obtain the owner's agreement that the owner will not make any claims against the EPC contractor's subcontractors. Once the owner sues a subcontractor, the subcontractor is likely, in turn, to sue the EPC contractor if the EPC contractor caused or contributed to the problem. Thus, the waivers that the EPC contractor solicited from the owner may not serve their purpose of protecting the EPC contractor from "uncapped" liability if the owner can create an "indirect" suit for which the EPC contractor may be liable.

No Excuse of Performance

In all cases, the EPC contract should make clear that what is being limited is the EPC contractor's monetary liability to the owner, not the EPC contractor's liability to complete the facility and the work. Liability limitations are not intended to excuse the EPC contractor's performance of any of its obligations under the EPC contract should its continued performance become unprofitable.

Contract Expiration

In most jurisdictions in the United States, contracts need not contain an expiration date. If a contract does not contain an expiration date, generally, in most jurisdictions in the United States, the contract will be deemed to continue for a reasonable period under the circumstances unless it is clear that the parties intended otherwise. Given that an EPC contract will typically contain a fixed warranty period but indemnification provisions that should survive beyond the expiration of the warranty period (see Chapter 23), it is to the owner's advantage that the EPC contract contain no stated expiration, or, if it does contain an expiration date, it should be one that occurs well after termination of the warranty period.

For instance, the equipment and control room in a facility is likely to employ patents and copyrights that are usually valid for decades, and to the extent that the EPC contractor has infringed upon a party's intellectual property rights, the owner should want protection long after the warranty period expires. If the EPC contractor does, however, insist upon a termination date, the owner may want to propose a compromise such as a reasonable period but in no case shorter than the period under the applicable statute of repose or term of any licensed technology used in the facility.

Notes

1 See *American Family Mut. Ins. Co. v. Pleasant Co.*, 268 Wis. 2d 16 (2004).
2 English law does not distinguish between negligence (sometimes referred to as "simple" negligence in the United States) and "gross" negligence.
3 In fact, in New York, even limitations on the "negligence" of a contractor are unenforceable (see Title 3 [General Obligations] of the Consolidated Laws of New York State). Also see *Union Carbide Corporation v. Siemens Westinghouse Power Corporation*, 2001 WL 1506005 (S.D.N.Y. Nov. 26, 2001), noting that limitations can be disregarded if a party has acted in bad faith. Also see *Abacus Federal Savings Bank, Appellant v. ADT Security Services, Inc., et al, Respondents, et al., Defendants*, No. 33, Court of Appeals of New York, 18 N.Y.3d 675; 967 N.E. 2d 666; 944 N.Y.S. 2d 443; 2012 N.Y. LEXIS 504; 2012 NY Slip Op. 2120.
4 *Hadley v. Baxendale*, 9 Exch 341, 156 Eng. Rep 145 (1854).

Chapter 21

Payment for the Work

The EPC contract should make clear that the EPC contractor will be paid a fixed price for its performance of the work and this price will not be subject to change unless the EPC contractor is expressly entitled to an equitable price adjustment by the terms of the EPC contract. It is advisable that the EPC contract actually contain a provision whereby the EPC contractor waives all claims against the owner for any additional compensation or for any damages that the EPC contractor incurs in connection with the EPC contractor's performance of the work so that there will be no doubt that the contractor is taking the risk that circumstances may arise that impact the cost of the work and the EPC contractor is putting its profit margin at risk.

Unlike general contracting, in which profit margins and "markups" are locked in, the EPC contractor is undertaking work at its own economic peril or benefit. That is why EPC work usually commands a premium price over general contracting work—because fixed price work is financially risky if unexpected circumstances arise. Of course, any savings that result from the EPC contractor's efficient performance will be kept by the EPC contractor. Thus, the EPC contractor has a great incentive to work efficaciously but if problems arise that cannot be attributed to *force majeure* or the owner's failure to comply with the owner's obligations, the EPC contractor can quickly find itself in the situation where its profit is wiped out and it is spending its own money to complete the owner's project and also, potentially, may be obliged to pay liquidated damages to the owner for delays in completing the project or the project's poor performance once it is built. Clearly, the EPC contractor must be sure it has included a reasonable contingency margin on top of costs and profit in case unanticipated events occur. Once the EPC contractor has spent its contingency margin, the only way it can preserve its profit margin is by shifting unexpected costs to the owner. This shift is precisely what EPC contracting is designed to prevent, except for a few specifically delineated classes of events such as those directly attributable to the owner and *force majeure*. Bearing all this in mind, as was discussed in Chapter 4, before signing the EPC contract it is advisable for the owner to review and evaluate the methodology that the EPC contractor used to arrive at the EPC contract price, including the EPC contractor's cost estimates for equipment, materials, labor and contingency. If it appears that the EPC contractor has planned to purchase "bargain basement" priced components, assumed low labor rates, or shaved its contingency and profit margins to the bare minimum, the owner should not be surprised when the change order requests begin to roll in and shoddy work begins to appear. In negotiation of the EPC contract price, the owner must balance its own desire for a low project capital cost against the necessity of certainty in the context of the EPC contractor completing the project on time and in accordance with the technical requirements of the EPC contract.

Exclusions and Inclusions

Taxes

The EPC contract should specify exactly which, if any, items are to be excluded from the EPC contract price. Often, in the interest of clarity, the EPC contract will clearly enumerate by way of example which types of taxes and duties are to be paid by the owner, such as sales tax on the EPC contract price, and which types of taxes will be the responsibility of the EPC contractor, such as export duties on components. In many cases, usually the owner, but occasionally the EPC contractor, will be exempt from certain taxes and will provide evidence of this exemption to the other party so that the concerned party can refrain from paying taxing authorities.

Because the owner will be required to depreciate its facility and equipment for taxation and accounting purposes once the facility is placed into service, the owner should request that the EPC contractor provide a description of the cost and type of major items of the facility so the owner can determine the proper depreciation classification and rate for each depreciable item. The EPC contract should also provide that the consequences of any change in tax law or rates will be borne by the party responsible for such taxes under the EPC contract and will not entitle the responsible party to relief or compensation if any such event occurs.

Spare Parts

The EPC contract should also specify that items such as spare parts and tools that are not physically integrated into the facility are included in the EPC contract price. The functional specification should include a list of exactly what additional items the EPC contractor will provide in connection with the work. It is also worthwhile to ask the EPC contractor to provide a list of spare parts that the EPC contractor and its vendors believe are useful to keep on site to reduce outage durations in the event of equipment failures. Purchasing spare parts during construction will generally also provide a hedge against rising prices for parts. Usually, the owner will be able to obtain better prices for spare parts if they are ordered at the same time as the EPC contractor places its initial order with a vendor (rather than waiting until a part is needed later during operation of the facility). Thus, during negotiation of the EPC contract, the owner should determine which parts the owner (or its operator) deems to be strategic and worthy of ordering at the outset of the project rather than paying an unnecessary premium for the part when it is urgently needed for service. Also, if the owner is financing its project, it may be better to include the cost of spare parts in the EPC contract price that is being financed by lenders and therefore may increase the owner's financial return on its project instead of using its own money to purchase the parts during operations.

"Take Out" Prices

Sometimes the owner will ask for a "take out" price for certain items if the owner has not yet decided whether or not it wants to include these items in the EPC contractor's work. An example might be a water intake facility that will have to be constructed on a river. The owner might execute the EPC contract while still in negotiations with the river authority because the river authority itself elects to build the intake structure. If the owner and the river authority cannot agree upon terms, the owner will require the EPC contractor to build the intake and structure. If a "take out" price has been included in the EPC contract and the owner does in fact reach an agreement with the river authority on the intake facility's construction, the owner can simply elect to remove the intake facility from the EPC contractor's scope of work and a corresponding and predetermined deduction to the EPC contract price can be made. It is also important to predetermine the price deduction for a take out item rather than leave it to the EPC contractor to calculate the take out deduction from the EPC contract price at the time the owner requests the "take out." This is true because the EPC contractor might be inclined to minimize the deduction if the EPC contractor has already encountered overruns on other aspects of the project by the time the "take out" is requested.

Obviously, another (but probably less desirable) way to handle this issue is giving the owner the option to include additional work in the EPC contract scope and price. Thus, in the case of a power plant, an owner might want the option to require the EPC contractor to build a temporary or permanent bypass stack for a combined cycle power plant so that it can run in simple cycle mode until ready to run in combined cycle mode. Typically, partially constructed power plants are often run in simple cycle during periods when electricity demand is high, such as in summer. The facility's temporary stack is knocked down when the plant is converted to combined cycle (usually over the winter) if not enough space is available for the equipment that remains to be installed. In most cases, it is to the owner's advantage to include everything that might be necessary in the EPC contract price and then make subsequent but pre-agreed-upon deductions if the owner decides an item is not necessary or desirable. If, instead, the EPC contractor has provided a list of options and their pricing, the EPC contractor will always have some leeway to argue that circumstances have changed since the option price was calculated. Therefore, an increase in price or a delay in schedule is necessary in connection with the owner's exercise of the option. It is probably more difficult for the EPC contractor to argue that an agreed-upon reduction to the EPC contract price should no longer be valid as written. Although this distinction may not be so significant in the context of equipment supply, it can prove important in the supply of items requiring extensive manpower and construction.

Currency Fluctuation

EPC contractors will often "quote" their EPC price in separate components according to different currencies (often to reflect and attempt to protect against the EPC contractor's own risk that many local subcontractors will often only accept the currency of their own jurisdiction). In the case that the EPC contractor will not execute the EPC contract calling for payment in a single currency, the owner may desire (or its lenders may require) that the owner enter into currency hedges to protect the owner from changes in currency exchange rates so that the owner eliminates the risk that it could require more funds to complete the project if currency exchange rates change.

Guaranteed Maximum Pricing and Risk Pools

While this practice is more typical in general contracting, occasionally an EPC contract may not set a fixed price. Instead, the owner might be charged on a "time and materials" basis with the understanding that the EPC contractor will not charge the owner more than a "guaranteed maximum" price (GMP) irrespective of the ultimate cost of the work. In such cases, it is also usual to establish targets and incentives (often referred to as "risk pools") for the EPC contractor to try to complete the project for less than the GMP. Several methods can be employed, but it is typical that a target price is set and the owner will pay only a (declining) percentage of the costs incurred above the target until the owner's spending has reached the GMP. This sharing of costs with the EPC contractor once the target has been exceeded serves to deter the EPC contractor from incurring unnecessary costs because such costs will not be borne solely by the owner. Using a GMP often complicates invoicing issues and is, therefore, not typical on large-scale projects (but may be used for a particular aspect of a project [such as tunneling work in the case of a hydroelectric project or subway]).

Incoterms

Since 1936, the International Chamber of Commerce (ICC) has published rules (Incoterms 2010 is the latest version) for the interpretation of 13 international commercial trade terms commonly used by buyers and sellers with regard to responsibilities for clearances, taxes, duties, insurance and risk of loss, but not ownership rights or relief from liability. While these "Incoterms" are useful for the sale of goods, they should not be used in the context of EPC contracts because EPC contracts call for responsibilities far beyond delivery, such as installation and testing. Including Incoterms can lead only to confusion and can actually conflict with many other provisions of the EPC contract, such as those relating to insurance and risk of loss. For instance, if the owner and the EPC contractor have agreed that the owner will be responsible for payment of import duties and have chosen to make an abbreviated reference to this concept in the EPC contract by agreeing to the Incoterm "DDU" (duty delivery unpaid) instead of describing the responsibilities of the parties, the owner might unwittingly shift the risk of loss for goods at the site from the EPC contractor to itself because that concept is implied by the Incoterm "DDU."

Rising Prices?

As was mentioned in Chapter 5, once the owner executes the EPC contract and has locked in the EPC contract price, if the owner has a long-term offtake or use agreement in place, the profitability of the owner's project will essentially be set unless the owner is financing its project and has success in obtaining better financing terms than expected (such as more debt, lower interest rates or a longer term) or in operating the facility more efficiently than expected.

As a development matter, the owner must determine whether or not postponement of the EPC contract's execution is likely to have a beneficial or deleterious effect on the EPC contract price. If the project is complicated and located in an area where EPC contractors are busy, the EPC contract price will probably tend to rise. Therefore, the owner should execute the EPC contract as soon as possible. Conversely, if the construction market in the area of the project is depressed and there are too many EPC contractors pursuing too few projects, it could be to the owner's advantage to let the force of competition drive the EPC price down over time and delay its signing.

Letters of Intent

If the owner believes that the EPC contract price is likely to rise over time and, therefore, wants to execute the contract as early in the process as possible, it is in the owner's interest to ask the EPC contractor to hold the EPC contract price "firm" for as long as possible. To do this, the owner could sign a letter of intent or memorandum of understanding with the EPC contractor. (Part A of Volume II contains a sample letter of intent.) Some letters of intent outline the material terms that the EPC contract will contain and memorialize. Other letters of intent merely reflect the agreement that the owner and the EPC contractor will cooperate in good faith for a limited period (often on an exclusive basis) in order to try to agree upon terms of the EPC contract. Such a letter of intent is essentially an agreement to agree and usually, in U.S. jurisdictions, will not bind a party to enter into a contract.[1]

Except to lock in an EPC contract price or price range, a letter of intent is of little value to the owner. The owner must make sure that the owner preserves its options to work with other EPC contractors if a deal cannot be struck. The owner must also make clear that any work product produced during the contract negotiations between the EPC contractor and the owner can be used by the owner in connection with the project and even by another EPC contractor—so the owner is not trapped into wasting time and money repeating preliminary scoping and engineering work.

If the owner believes that a letter of intent is not a profitable use of time and resources, or a letter of intent has, in fact, been executed but contract negotiations have bogged down, or the owner has strategic reasons why it does not want to execute the EPC contract, the owner could consider entering into a limited "notice to proceed" agreement (as was discussed in Chapter 5) as a method to hold the EPC contract price.

If the owner has exhausted the alternatives above or determines that the best approach is, in fact, to enter into the EPC contract, the owner should negotiate a provision that will permit the owner to delay issuing notice to proceed with the work to the EPC contractor for some predetermined period during which window the EPC contract price will not be subject to change. Most EPC contractors usually will accommodate some period (often up to 90 days) in which the EPC contract price will not be subject to change if the owner issues the notice to proceed within this period. It is also to the owner's advantage for the EPC contract to allow that if notice to proceed is not given during this period, the owner can still give notice at a later date but some (or all) of the EPC contract price will begin to escalate until notice to proceed with the work is given to the EPC contractor by the owner. This escalation can be based on any index or formula the parties agree upon, but most typically escalation is based on exchange rates and price indices.

For example, consider a coal-fired plant being built in the Philippines. The EPC contract price may be payable in U.S. dollars, but the steam turbines are being sourced from France. Therefore, if 25 percent of the EPC contract price is attributable to the steam turbines, so 25 percent of the EPC contract price might escalate by the French producer price index. Then, this escalation component would be converted into dollars at the exchange rate of euros for dollars on the date that notice to proceed is given to the EPC contractor. The remaining 75 percent of the EPC price might be escalated by the Philippine producer price index and similarly converted to dollars on the date that notice to proceed is given.

Finally, even if the EPC contractor insists that an outside date for notice to proceed must be set because after that date use of an escalation formula as a proxy for actual escalation is too risky for the EPC contractor, the EPC contract should provide that notice to proceed can still be given after the outside date but that the EPC contractor will be entitled to an equitable adjustment to the EPC contract price to cover the EPC contractor's increased costs during this waiting period. In this way, even though the owner will not have been able to keep the EPC contract price firm, the EPC contractor will not be in a position to renegotiate any terms of the EPC contract other than the EPC contract price (and possibly schedule). This is important because it will preclude the EPC contractor from changing the risk profile of the project for the owner by attempting to revisit agreements on terms contained in the EPC contract, such as those concerning target performance levels of the facility and warranties. If the owner cannot freeze the EPC contract price and project schedule, at least it will have been successful in freezing the balance of its commercial deal with the EPC contractor.

Milestone Payments

Unlike a general contractor that will usually be paid progress payments in accordance with some interval (typically a month), the EPC contractor will be paid in allotted increments as it completes itemized tasks, often referred to as "milestones." At the time the EPC contract is executed, the EPC contractor and the owner will have agreed upon a list of actions whose successful completion will entitle the EPC contractor to an agreed-upon portion of the EPC contract price. Sometimes this proportion is expressed as a percentage of the EPC contract price and sometimes it is stated as an amount. Although the protocols are equivalent, stating the payment due as a percentage of the EPC contract price can become confusing if change orders to the EPC contract price are subsequently executed. Stating the portion of the EPC contract price associated with a milestone event can avoid this ambiguity.

Timing

Payments of the EPC contract price should correspond to the completion of tasks—not calendar date (see Figure 21.1). It may be helpful for the owner's cash flow projections if the EPC contractor includes in the EPC contract the date on which the EPC contractor expects to achieve each milestone event, but the payment schedule in the contract should make clear that calendar dates are provided for projection purposes only and bear no significance as to when payments are due. Once the EPC contractor completes a referenced activity, it will be entitled to a milestone payment irrespective of the calendar date upon which the work was completed. Consequently, if the owner has not paid careful attention to the milestone events themselves, the owner could find itself owing the EPC contractor a lot more money than expected at any given point in the project's development. For instance, suppose the EPC contract milestone payment schedule requires a 10 percent milestone payment when the owner gives the EPC contractor notice to proceed with the work and then further milestone payments as the EPC contractor awards each subcontract for major equipment. Typically, EPC contractors award equipment orders to vendors by means of a letter notification to a vendor so that work can be commenced by the vendor and the project schedule not be jeopardized. The actual subcontract itself is usually not executed until sometime later after the subcontract has been negotiated between the subcontractor and EPC contractor. This delayed execution of subcontracts is generally not a big risk for an EPC contractor because most of its subcontracts are typically in standardized forms and sometimes simply preprinted purchase orders.

It is entirely possible (and not uncommon) for the EPC contractor to request payment for all these subcontract award milestones within days of the receipt of notice to proceed by the EPC contractor, so an unwary owner could suffer an unexpectedly premature cash outflow to the EPC contractor. This problem can be avoided in a number of ways. First, the owner could evaluate the milestone events themselves to determine whether or not they are appropriate and true signals of progression of the work and select other milestones if they are not. Second, the owner could specify that certain milestones will not be paid before a certain period has elapsed since notice to proceed has been given irrespective of when they are achieved. Third, but less effective than the first two approaches, the owner could specify that invoices for milestones may not be submitted more often than once in any specified period, usually once each calendar month. It is typical that this would be specified anyway in the EPC contract because if the owner has financed its project, its lenders generally permit the owner to draw loans only once per month for administrative reasons, usually because the lenders themselves must borrow the money that they will lend to the owner.

Finally, it is typical for EPC contractors to insist that they be kept "cash neutral" and that they be paid in advance for their monthly expenses so that they are not out-of-pocket if the owner ceases to pay them for some reason. Thus, the EPC contractor may insist on billing the owner for milestones that the EPC contractor expects to achieve in the coming month and at the end of the month the owner will determine which milestones were achieved and make a reconciliation against the upcoming month's payment, if necessary. In the alternative, milestone payments may be "inflated" to cover upcoming work costs and not just reflect the cost of the work associated with the milestone whose payment is due.

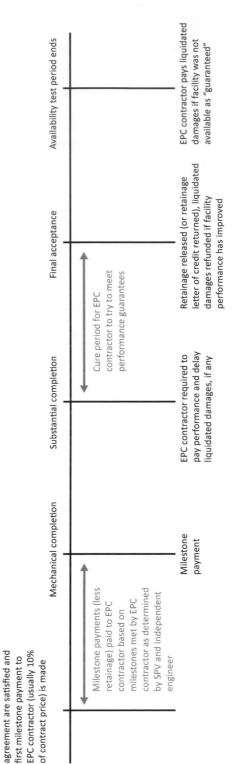

Figure 21.1 Typical project payment structure.

The Right to Verify Progress

The EPC contract will provide that the owner and its representatives (including the lenders and their representatives if the owner has financed its project) may inspect the EPC contractor's work prior to the owner's payment of any milestone so that it can be determined whether or not the work corresponding to the milestone payment requested has, in fact, been completed in accordance with requirements of the EPC contract. While this provision seems benign, the owner must make sure these inspections will not have the effect of precluding the owner from raising claims in future that the EPC contractor's work is not in accordance with the EPC contract. Therefore, it is crucial that the EPC contract state that any inspection of, or payment for, any work shall not constitute the owner's acceptance of the quality of the work or be deemed to be a waiver of any of the owner's rights to raise any claims (warranty or otherwise) whatsoever in the future regarding the EPC contractor's performance.

Retainage

The notion of not paying a contractor in full until the contractor has completed its work properly probably dates back as far as the concept of paying someone to build something. The retention of a portion of payment is generally known as "retainage." In EPC contracts, a portion of each milestone payment is usually withheld as retainage, typically between 4 and 10 percent of each milestone payment. Retainage often corresponds to the EPC contractor's profit margin. Thus, the net amount of each milestone payment by the owner to the EPC contractor will usually be sufficient to allow the contractor to cover its costs of performing the work so that the EPC contractor will not have to borrow money, or use its own funds, to finance the cost of the work, but the EPC contractor (theoretically, at least) will not be able to earn any of its profit yet because (again theoretically) there will be no payment by the owner to the EPC contractor in excess of the EPC contractor's costs until the project is complete. Although this may sound like science, even if the EPC contractor has disclosed its anticipated profit margin to the owner before signing the EPC contract (as is supposed to be the case in the "open book" negotiation method discussed in Chapter 4), the chances are good that the EPC contractor is hoping to exceed this margin. On the other hand, unexpected overruns can quickly wipe out all the EPC contractor's expected profits.

Retainage is usually withheld until substantial completion or provisional acceptance (or sometimes even until final acceptance, if the owner has driven a hard bargain in the EPC contract). Since the project may produce little free cash for the EPC contractor to use for other endeavors during the project's lifecycle (which could be years), if retainage is withheld, the EPC contractor may want to compensate for this cash deprivation. The EPC contractor may attempt to inflate the EPC price to try to compensate for its opportunity cost of being unable to deploy the retainage. The owner will obviously resist this strategy if the owner is aware of it. Typically, the owner actually will pay the "retainage" to the EPC contractor with each milestone payment so long as the EPC contractor has delivered a letter of credit in the amount of the retainage that should have been withheld by the owner. This letter of credit will be drawable as retainage if the owner encounters a problem for which the EPC contractor is responsible under the EPC contract. In this arrangement, the EPC contractor receives its profit with each milestone payment and the owner remains protected in the sense that the owner can have its cash retainage returned to it at any time by drawing upon the letter of credit. In fact, in the United States, in some sense, this arrangement is actually better for the owner than withholding the EPC contractor's profit because if the EPC contractor has made a bankruptcy filing, the owner could conceivably be restrained by the bankruptcy court from applying the cash retainage that the owner is withholding because other creditors of the EPC contractor could argue that this retainage is part of the EPC contractor's estate and, therefore, should be used for the benefit of all the EPC contractor's creditors and not just the owner. On the other hand, if the owner is holding a letter of credit, since the letter of credit is payable by its issuer (a bank) and not the EPC contractor, a court will not be able to curtail the owner's ability to draw upon the letter of credit (and apply the funds as necessary) because they are not the EPC contractor's funds. Thus, the issuer of the letter of credit, rather than the owner, will be left "holding the bag" with a reimbursement claim in the EPC contractor's bankruptcy proceedings. Given the number of large EPC contractors that have filed for bankruptcy protection over the years, this matter may be worth consideration. Of course, there are banking fees involved in posting and maintaining a letter of credit, which will vary based upon the creditworthiness of the EPC contractor. The EPC contract should make clear which party will bear these fees and typically it is the EPC contractor. (See Chapters 13 and 19 for further discussion of letters of credit and bankruptcy principles.)

There are other regimes that EPC contractors employ in EPC contracts to prevent potential cash drains. For instance, the retainage percentage of each milestone payment may begin relatively low (such as 2 percent) and increase over time (up to around 10 percent) so that more money is withheld later in the project lifecycle. Or, no retainage at all may be withheld but the milestone payment associated with substantial completion (when most retainage is usually released) may be greater than normal (for example, 20 percent of the EPC contract price). Typically, the milestone payment due at substantial completion is between 7 and 10 percent of the EPC contract price. The next, and last, milestone payment due upon final acceptance is often between 2 and 5 percent of the contract price. Thus, the milestone payment due upon substantial completion or provisional acceptance could be amplified from 10 to 20 percent of the EPC contract price to make sure that the owner is withholding a sufficient amount of the EPC contract price so that the owner will have adequate funds available to take any remedial action that may be necessary.

It is important not to confuse the purpose of retainage (or a retainage letter of credit) with the purpose of a "performance" letter of credit for liquidated damages (discussed in Chapter 19). Retainage is withheld so that the owner will have funds to complete work that should have been (but is not) completed by the EPC contractor. Retainage ensures that the owner will not have to make an expenditure in addition to the EPC contract price and then be forced to try to reclaim this additional expenditure from the EPC contractor, who may be bankrupt or simply unresponsive.

Suppose a selective catalytic reduction unit (SCR) that the EPC contractor has installed to reduce nitrous oxide emissions uses excessive amounts of ammonia, the catalyst in the reduction process. Suppose further that the EPC contractor has tried unsuccessfully to fix the problem. If the owner is holding cash retainage, the owner can use this retainage to hire another contractor to try to fix the problem without having to outlay the owner's own money and then hope to recoup this money from the EPC contractor. This would be the case if there had been no retainage withheld and the owner had already paid most of the EPC contract price to the EPC contractor. Essentially, retainage can prevent the owner from paying twice for the same item (and paying legal fees should the EPC contractor refuse to reimburse the owner for the owner's efforts). For instance, the EPC contractor could claim that the owner has not been reasonable in incurring remedial costs in its attempt to fix the problem or claim that there is no problem but that the owner has changed its mind and now desires a level of performance that it did not originally specify that it wanted. By contrast, the performance letter of credit is in place to secure payment of money that the EPC contractor may owe the owner for delays or retarded performance, not money that the owner might need as a result of the EPC contractor not completing the work. For instance, if in addition to the SCR problem the EPC contractor encountered a transportation delay in the delivery of the SCR and, therefore, the EPC contractor did not complete the facility by the guaranteed completion date that the EPC contractor promised in the EPC contract, it would also owe the owner liquidated damages for the delay in completion of the facility. If the EPC contractor did not pay the liquidated damages to the owner (perhaps because it was claiming that the delay was excused by *force majeure*) and the owner applied the cash retainage that the owner was withholding to satisfy the unpaid liquidated damages, the owner might have no retainage money left to hire a contractor to fix the SCR.

Thus, if liquidated damages are owed and not paid by the EPC contractor, the owner should draw upon the performance letter of credit and not apply the retainage that the owner is holding. In fact, if the offtaker or user has assessed liquidated damages against the owner for the facility's delay in becoming operational and/or retarded performance, the owner could face a serious cash crunch if the owner were to try to use retainage (which will probably be insufficient in any event) to fund all these sudden cash requirements. A performance letter of credit will help prevent the owner from being caught in the position of using its own cash when problems arise and then trying to recoup its money from the EPC contractor.

Release of Retainage

Once the EPC contractor achieves substantial completion or provisional acceptance, only punch list work will remain to be done. At that point, the owner will be satisfied that the EPC contractor has performed all significant work. Therefore, most of the retainage can be released to the EPC contractor. Of course, the owner will typically retain about 150 or 200 percent of the value of the items on the punch list in case the EPC contractor does not complete these items by the final acceptance date that the EPC contractor has promised in the EPC contract. Rarely, but occasionally, retainage is withheld throughout the warranty period because, if a problem arises during the warranty period, the owner will usually not have easy access to the EPC contractor's cash to secure the EPC contractor's obligations unless retainage is withheld or a warranty letter of credit has been posted (which is not typical either). Consequently, it is important to engage an EPC contractor that is likely to be around in the future to stand by its work—a particularly difficult problem if a consortium was formed by contractors for the sole purpose of carrying out the EPC contract and parent guaranties were not obtained.

Payments Withheld

Distinct from retainage to cover work yet to be performed by the EPC contractor is the practice of withholding payment once a specific issue arises that may affect the owner's liability but not performance of the facility. To deal with these circumstances, the owner usually negotiates the right to withhold payment of a portion of the EPC contract price in an amount sufficient to hold a cash reserve against any potential liability the owner might have to third parties. There can be many examples of this. For instance, suppose the owner along with the EPC contractor is sued by a neighboring landowner that claims that work conducted at night is loud and has created a public nuisance and claims damages for emotional harm and also seeks an injunction to stop the contractor from working at night. While the suit may not be successful and insurance generally would cover this kind of loss (although many policies cover only physical injury and not emotional distress), the owner might want to begin withholding payments from the EPC contractor to build a fund to cover the owner's legal costs or in case the insurer does not honor a claim or the policy contains a deductible. The owner will not have to seek reimbursement from the EPC contractor if the owner has withheld funds. Withholding funds is obviously less burdensome than pursuing recovery from the EPC contractor under the indemnification provisions of the EPC contract (to be discussed in Chapter 23), especially if the EPC contract failed to proscribe the EPC contractor's working at night.

Other examples of circumstances in which withholding may be prudent arise with subcontractors. Suppose a construction subcontractor believes it is entitled to an equitable price adjustment of its contract price based on *force majeure* because heavy rains have delayed its work for two weeks and generated unexpected expenses in order for the subcontractor to keep construction on schedule. The unhappy subcontractor now requests an additional $200,000 in overtime labor costs from the EPC contractor. What if the EPC contractor rejects the claim of *force majeure* and refuses to pay the additional $200,000? As a retaliatory offensive tactic, which will certainly get the attention of all project participants, if there is a state statute that allows a workman that has not been paid to file a lien against property into which its services have been incorporated, the subcontractor could file a lien against the facility for $200,000. As a practical matter, the owner might need to satisfy this lien or risk foreclosure of the subcontractor's lien against its facility. To protect itself, and, if it has financed its project, to comply with requirements of its lenders (which usually provide that liens above a certain *de minimis* threshold be promptly extinguished or bonded) the owner will ask the EPC contractor to cause the lien to be lifted. If this is not done promptly, perhaps because the EPC contractor and its subcontractor may now be in the midst of a lawsuit, the owner has a few choices. First, the owner could withhold $200,000 from the next milestone payment due to the EPC contractor and wait to see how the dispute is resolved, so long as this approach does not violate terms of the owner's credit agreement with lenders if the owner has financed its project (which often require that liens be released within a certain number of days after they have attached to the owner's property). Second, the owner could simply pay the subcontractor the $200,000 so the subcontractor releases the lien and leave the EPC contractor to try to collect the $200,000 from the subcontractor. That approach will probably not be well received by the EPC contractor and is generally not a good idea. Even if the EPC contractor has finished the facility by the time this payment is made, the owner may still need the EPC contractor's (expedited) services during the warranty period, especially to avoid unnecessary outage time while warranty items are awaiting repair. So, it is probably best for the owner not to antagonize the EPC contractor by forcing the EPC contractor to try to recover money from an insouciant subcontractor. Finally, the owner could post a bond in the amount of the claim giving rise to the lien.

The EPC contract should provide that the EPC contractor will be responsible for costs incurred by the owner in taking any of the above courses of action.

Setoff

The right of "setoff" is slightly different from the right of withholding. In the case of "withholding," a party is making provision for a contingency that may or may not materialize (for example, the subcontractor's claim for *force majeure* discussed above may be resolved against the subcontractor). In a setoff, a party is merely reducing a payment that it is required to make by the amount of a payment required to be made to it by its counterparty. For example, if the EPC contractor achieves substantial completion or provisional acceptance two days late, the owner will owe the EPC contractor the entire milestone payment due at substantial completion or provisional acceptance and the EPC contractor will owe the owner delay liquidated damages for the two days of delay in achieving substantial completion or provisional acceptance. In this event, so long as the EPC contract provides for setoff, the owner can simply net the amount it owes to the EPC contractor against the two days of delay liquidated damages and pay the balance to the EPC contractor. Even if the EPC contract does not provide for setoff, a right of setoff usually exists under common law in most U.S. jurisdictions, but it is best not to rely on this right because there may be many defenses to setoff available to the party whose debt is being set off. While setoff is generally a non-controversial topic, in rare but serious cases, such as patent infringement disputes, the stakes and legal fees can be astronomical and whether a party is able to set off can have serious cash flow consequences if it cannot. (Loan agreements will almost always preclude a borrower from setting off against amounts it has been lent.)

Disputed Payments

With regard to payments in dispute, the owner should either retain the funds in dispute or agree to deposit them with an escrow agent pending the resolution of the dispute. Sometimes an EPC contractor will suggest that one-half of the disputed amount be turned over to it while the dispute is in progress. While this sounds fair, on reflection it can be a potentially costly compromise for the owner. For example, if a *force majeure* event occurs and the owner believes the EPC contractor should be entitled to $100,000 of additional costs and the EPC contractor believes it is entitled to $300,000 of additional costs, must the owner turn over to the EPC contractor half of $100,000, half of $300,000 or half of $200,000 (the difference between the $300,000 claimed by the EPC contractor and the $100,000 estimated by the owner)? Since it may not be clear what "half" really means in this context, it is probably preferable for the owner to agree to pay interest (at an agreed-upon rate) on any disputed amount that the owner has retained that is ultimately determined to be due to the EPC contractor. Naturally, a similar result would be obtained if funds were to be deposited with an escrow agent and those funds earned interest while under the escrow agent's control.

Unfortunately, an escrow arrangement has the disadvantage of requiring the finding of an escrow agent, negotiation of an agreement between the owner, EPC contractor and escrow agent, and the payment of fees to the escrow agent. Usually, this is not done in advance because it is deemed wasteful of the time and effort needed to negotiate an agreement (especially since disputes of a magnitude to justify an escrow arrangement are relatively rare). This is unfortunate because it is unlikely that the parties will be on cordial terms when the escrow arrangement needs to be negotiated because a dispute has arisen. Furthermore, the weak underbelly of all escrow arrangements is the escrow agent's own temerity (for personal liability reasons) in disbursing funds to a party. If a party believes the escrow agent is about to disburse funds improperly under an escrow agreement because conditions to the escrow release have not been met, the aggrieved party need only sue—sometimes merely threaten to sue—the escrow agent and the agent is likely to stop in its tracks until it receives directions from a court or arbitrator as to how to release the funds, which could take years. Thus, the true *raison d'être* for the escrow can sometimes be vitiated by the mere threat of a lawsuit.

Additionally, the EPC contract should contain a provision requiring that interest (at a high rate) should accrue on overpayments to the EPC contractor by the owner and on late payments made by the owner to the EPC contractor. This will discourage the EPC contractor from overcharging the owner by submitting payments for milestones that have not been fully met and will discourage the owner from using the EPC contractor as the owner's working capital lender by stringing out payments to the EPC contractor.

Payment Currency

Payments under an EPC contract should always be denominated in, or indexed to, a "hard" currency (like U.S. dollars, euros, yen, etc.) if local law prevents payments in foreign currencies. For instance, if the owner has set up an SPV in the jurisdiction of the project and the EPC contractor has also set up an SPV in the jurisdiction of the project to enter into the EPC contract, these companies might be required to transact business in the local currency. However, the owner should be cognizant of the fact that inflation and currency devaluation over a project's construction lifecycle have the potential to erode the value of the liquidated damages payable by the EPC contractor. Similarly, indemnity payments (which EPC contracts often provide) must be made in the currency in which they have been incurred by the indemnified party, and sometimes may not be reimbursed until years after the loss in question has been suffered by the indemnified party. In all of these cases, the EPC contract should provide a mechanism to index the payments (preferably to a "hard" currency) in order to try to compensate for potential timing mismatches.

Suspension of Milestone Payments

Occasionally, owners will request that the EPC contract protect the owner against being required to make payments to the EPC contractor if the EPC contractor is in default under the EPC contract. While this sounds like a good idea, the owner's attempt to include this provision is likely to irritate the EPC contractor, who will try to limit applicability of this "penalty" to projects in which an owner has financing and in which a default by the EPC contractor has the effect of stopping the flow of loan disbursements from lenders to the owner. Each time the owner borrows money under its loan agreement to pay the EPC contractor, the owner (and usually the EPC contractor) must certify the state of completion of the project to the owner's lenders and often even certify that no event has occurred that could have a material adverse effect on the project or the owner's ability to repay its loans. If the EPC contractor is in bankruptcy or is not working diligently, the lenders will typically negotiate the right not to advance funds under their credit agreement. Presumably, however, if this really were the case and the EPC contractor had ceased working, the owner would have much more serious problems and it is not realistic to believe that this provision would be of any help to the owner (see the discussion of "adequate assurance" in Chapter 13).

Payments for Unresolved Change Orders

As will be discussed in Chapter 24, the EPC contract should never permit the EPC contractor to slow or stop its work unless *force majeure* has occurred or the owner is not paying the EPC contractor. While an equitable adjustment, for whatever reason, is pending, the case is no different and the EPC contractor should continue to work. The EPC contractor should be paid for its work and should not be required to advance costs for a change in the work for which the EPC contractor is not responsible under the EPC contract.

Suppose, in the case of a power plant, that the EPC contract anticipates that bottom-dumping coal railroad cars will serve the coal stockpile because this type of railroad car can be unloaded quickly and the owner finds this design acceptable. But suppose that later, when the owner is in negotiation with its coal supplier, the owner is told by the coal supplier that bottom-dumping railcars are subject to pilferage while they are temporarily stopped in transit because itinerants will dump an entire railcar load to pilfer a few hundred pounds of coal and then leave the rest of the coal car load to waste on the tracks. The owner quickly concludes that side-tipping railcars are more appropriate for that particular country, and asks the EPC contractor to make an appropriate change in the design of the stockyard to accommodate side-tipping cars. While the EPC contractor may be happy to make this modification, it will cost money and time, which the contractor did not contemplate. Therefore, an equitable price adjustment will have to be made as a result of this owner-directed request. While the EPC contractor will commence the design work right away, suppose the EPC contractor requests a lump-sum change order of $600,000 but the owner believes this supplemental work would only cost $250,000 if the EPC contractor solicited competitive bids from different design firms instead of simply turning to the firm the EPC contractor had previously used and asking for a quote on the additional work (and then applying its own contingency and profit margin on top of this sole quote). To avoid a predicament like this, the EPC contract should always provide that the EPC contractor should act reasonably in incurring costs associated with change orders because these are costs for work that was not anticipated at the time the EPC contract was executed. Thus, the owner did not estimate the price of these costs before entering into the EPC contract. In the case of the railcars, it is in the owner's interest for the EPC contractor to commence the new design work as quickly as possible so that time is not lost and side-dumping railcars can be ordered and fabricated. Furthermore, since the owner itself will probably have little basis on which to dispute the EPC contractor's selection of a design firm, it is probably best for the owner to acknowledge that the EPC contractor's selection is proper but attempt to continue to negotiate the price of the change order related to the supplementary design work. During the time the change order is under negotiation, the EPC contractor will want some payment from the owner to cover the EPC contractor's costs of additional overhead, supervisors, and the design firm.

While most EPC contracts fail to address the issue of how the EPC contractor will be paid while a change order is being negotiated or disputed, as was discussed in Chapter 12, the EPC contract should provide that any work that is so-called "out of scope" work will be performed on a time and materials basis, ideally at rates set out in an exhibit to the EPC contract. It is probably not possible to include all items or specialties that might be required in a change order, but a list of typical expense items, such as skilled and unskilled labor rates and materials such as concrete and rebar, will help prevent misunderstandings. Sometimes, the EPC contract is not as clear as it should be and merely provides that the owner will pay the EPC contractor the "undisputed" portion of any pending change order until the dispute is resolved. Unfortunately, it is not intuitively clear what is meant by an "undisputed" portion in the case of out-of-scope work. A separate dispute can arise over what "undisputed" means if the EPC contractor merely gives a lump-sum estimate that has no detailed cost breakdown. It is also desirable that the EPC contract make clear how frequently the EPC contractor will be paid for work until a change order is executed (such as monthly progress payments in arrears or upon completion of "sub-activities").

Special Payment Arrangement for Export Credit Agencies

As mentioned in Chapter 9, many government export credit agencies have programs to encourage the sale of goods and services from their respective countries. Consequently, proceeds of loans they make to an owner usually must be disbursed directly to the supplier of goods being financed by the export credit agency and not another party like the EPC contractor. Thus, if the owner is building a power plant in India and has retained a local contractor to design and construct the plant utilizing General Electric combustion turbine technology manufactured and assembled in Japan under a license to Mitsubishi Heavy Industries (MHI), the Japanese export credit agency, Japanese Bank for International Cooperation (JBIC), may require that, as the owner borrows from JBIC to finance payment for the turbines, the loan proceeds should be paid directly to MHI rather than to the EPC contractor (who in turn would normally pay MHI, the turbine supplier). Consequently, the EPC contract in these cases must provide that the owner can make payments directly to the relevant subcontractor and these payments will be credited against the EPC contract price.

In addition, most export credit agencies have more flexible programs under which they will agree to finance most of the cost of an entire project so long as the work carried out in connection with the project contains some minimum level of "content" from their home country. An export credit agency might agree to finance an owner's project if, for example, 75 percent of the project's content of goods and services originate in the export credit agency's own country. Alternately, some export credit agencies will finance a project so long as its owners are from its home country regardless of where the goods and services originate. In the most typical structures, the export agency will lend only a portion of the funds that the owner needs to borrow to build a project. The remaining portion of funds will be lent by commercial banks, usually a combination of local and foreign institutions, whose loans will often be guaranteed against political risks such as expropriation, but not commercial risks, such as poor plant operation, by the export credit agency. (See Chapter 23 for a discussion of political risk insurance.)

In all these cases, the EPC contract should provide that the EPC contractor will comply with the guidelines and requirements of the involved export agency (environmental guidelines, fair labor practices, etc.) so as not to jeopardize the owner's eligibility for the export agency's financing. Since the loss of eligibility for financing could have a catastrophic effect on the owner and its commercial banks, it is not uncommon for the EPC contract to provide that if the EPC contractor causes the owner to lose its export credit agency financing, the EPC contractor will (in effect) agree to convert the EPC contract price into a loan on the same rates and terms as the "lost" export credit agency financing by postponing its payment terms to correspond with those credit terms of the export credit agency's loan. While this seems like a draconian remedy, it could be the only hope the owner has for completing its project if an export agency's guidelines have been violated.

Termination Payments

To deal with catastrophic events if they do arise, the owner will reserve the right to terminate some, or all, of the EPC contractor's work at the owner's convenience. In such cases, the owner will typically agree to pay the EPC contractor for the costs the EPC contractor incurs in ceasing the portion of the work being terminated by the owner, including costs of terminating subcontracts, which may mean forfeiting down payments and demobilizing workers. The owner should also pay for any work that has been performed by the EPC contractor but has not yet been invoiced because the milestones that cover this work have not yet been, and now may never be, met. The owner should also be obligated to deliver to the EPC contractor any retainage it is holding that is associated with work that has been terminated. As was discussed in Chapter 15, whether or not the EPC contractor should be entitled to any expected profit in connection with a termination of work is a complicated issue. As was also discussed in Chapter 15, to avoid wasting negotiating time discussing the philosophy of compensation for lost profits, it is probably advisable for the owner and the EPC contractor to agree upon a termination schedule that sets forth the termination payment that will be due to the EPC contractor based upon what milestones have been achieved by the EPC contractor by the date of termination or upon how many days have elapsed between the owner's issuance of the notice to proceed and the notice of termination.

Realistically, if the owner is availing itself of its termination rights, it is likely that it is also going to be availing itself of the protections of the U.S. Federal Bankruptcy Code. Therefore, the owner will not be very concerned about the amount of money it will owe the EPC contractor, whose claim is unlikely to have any special priority over the owner's other creditors in its bankruptcy proceeding. In fact, if the owner has financed its project, the EPC contractor's claim will almost always be second in priority to the claims of the owner's lenders, which will almost always have been granted a first lien over all the owner's property as collateral for the repayment of their loans.

Security for the Owner's Payments of the Contract Price

These days, EPC contractors are more savvy about insisting upon the owner's ability to ensure that its payment obligations to the EPC contractor will be satisfied. At one time, owners could merely promise to pay the EPC contractor and, if the owner were financing its project, advance the argument that even if the owner did not, its lenders would pay the EPC contractor because the lenders would have no hope of repayment of their loans unless the facility was built. As was demonstrated in the U.S. power industry in the 1990s, this is not necessarily true. Lenders did not pay to complete projects being built in markets where power price predictions did not forecast that their debt would be recouped on approximately the same schedule that they were "banking" on. Consequently, it has become standard operating practice for EPC contractors to request a guaranty from the owner's ultimate parent entity (or even a letter of credit in the entire amount of the EPC contract in the case of unstable governments that have contracted for a project) in order to provide a backstop for the owner's obligations under the EPC contract. While it is unlikely that the owner will be able to avoid providing any credit support for its payment obligations, the owner should be able to limit the amount of any payment security that it does provide to either a specified percentage of the EPC contract price or a fixed amount, such as the sum of the next two or three upcoming milestone payments under the EPC contract. An EPC contractor will usually be comfortable with this limited credit support because the EPC contractor's goal is to stay out of the red and earn its profit.

The EPC contractor certainly can, and will, cease work to mitigate its damages if the owner fails to make payment as required by the EPC contract. The EPC contractor also understands that the owner is never really obligated to pay the entire EPC contract price until substantial completion or provisional acceptance is achieved because (as discussed above and in Chapter 15) the owner always has the right to terminate the EPC contract at any time for the owner's convenience. To address this scenario, a thoughtful EPC contractor will insist that any credit support provided by the owner (such as a guaranty from its parent entity or a letter of credit) should also cover the entire amount of any termination payment due to the EPC contractor if the owner terminates the EPC contract for convenience, because the EPC contractor will, also in turn, have to cancel all of its subcontracts and pay termination fees to its subcontractors. Given this concern, as was discussed above, the owner must understand what the termination payment is likely to encompass if a termination price schedule has not been agreed upon in advance.

Credit support of the EPC contract price is one of the few areas in which the owner's and, if the owner has financed its project, its lenders' interests vis-à-vis the EPC contractor are not generally aligned. This is true because, to the extent the parent entity of the owner has made payment on behalf of the owner to the EPC contractor, the EPC contractor's claim against the owner will be reduced by that amount. If the owner goes bankrupt, the lenders will be (marginally at least) in a better position than they would have been had the payment by the owner's parent not been made because the size of the claims against their borrower (the owner) will be reduced by the amount of the payment received by the EPC contractor from the owner's parent. Of course, the lenders' claims will be more likely to be satisfied than the claims of the EPC contractor anyway, because the lenders have taken collateral for their loans (the owner's facility and contracts) and the EPC contractor usually has taken no collateral from the owner to secure the owner's payment of the EPC price. In fact, quite the opposite. As was discussed, under terms of the EPC contract, the EPC contractor will usually have waived its rights to file liens against the owner's facility so that lenders will have no competition for their collateral (as discussed in Chapter 13).

Subrogation

Under common law in most U.S. jurisdictions, if a guarantor (such as the owner's parent entity) makes a payment under a guaranty to the beneficiary of the guaranty, the guarantor is subrogated to the beneficiary's claim against the original obligor (the owner in this case), which means that the guarantor can seek payment against the owner as if it were the EPC contractor. Typically, any party that gives a guaranty would have a written agreement with the obligor to the effect that the obligor will reimburse the guarantor (with interest) for payments made by the guarantor under the guaranty. However, if the owner has financed its project, its lenders will require the owner's parent guarantor to waive (or at least suspend) subrogation claims against the obligor (owner). The guarantor should resist lenders' requests that the guarantor permanently waive such a subrogation claim against the owner and suggest to the lenders that the guarantor is prepared to forbear from exercising its rights to pursue such a claim until the lenders have been paid in full (at which time the lenders should no longer be concerned). Sometimes, the lenders' lawyers will insert the word "irrevocably" before the words "paid in full" on the theory that if the lenders have to return any payments that they receive from the borrower (perhaps under the operation of the U.S. Federal Bankruptcy Code [as was discussed above]), they still do not want to compete with the guarantor in seeking funds from the owner. While it is probably not worth fighting about, the practical effect of adding this word "irrevocably" could create a permanent waiver of this right of subrogation because it is theoretically possible that some party may not be bound by the owner's bankruptcy discharge order if such party did not know about the owner's bankruptcy, and then the party could attack the validity of the discharge order and file new claims against the owner that could hypothetically force the lenders to return a payment that they received from the owner.

Note

1 See *Teachers Ins. & Annuity Assoc. v. Tribune Co.*, 1984 WL 645, *6 (S.D.N.Y. 1984) noting that: "A primary concern for courts is to avoid trapping parties into contractual obligations that they never intended. Ordinarily in contract negotiation, enforceable legal rights do not arise until either the expression of mutual consent to be bound, or some equivalent event that marks acceptance of an offer. Contractual liability, unlike tort liability, arises from an agreement to be bound (or in any event from the manifestation of consent). It is fundamental to contract law that mere participation in negotiations and discussions does not create any binding obligation, even if agreement is reached on all disputed terms. More is needed than agreement on each detail, which is overall agreement (or offer and acceptance) to enter into the binding contract."

Chapter 22

Title and Other Legal Matters

Reflecting development of the Uniform Commercial Code (UCC), which has been adopted in various forms by all 50 states of the United States, and the increased sophistication of commercial and financial transactions such as the pooling of credit card receivables and the swapping of commodities, the role of "title" in the United States to a good or service has diminished in importance. The last stronghold in the United States for the construct of "title" is probably real estate transactions. This is true because title often has little to do with connoting which party really has the economic and controlling interest in the property in question. Unfortunately, the concept of "title" is not really descriptive enough to deal with the intricacies of modern trade. Many other conventions have developed, such as the Incoterms referred to in Chapter 21, which are beginning to supplant the concept of "title."

In the United States, an important concept reflected in the UCC is the "security interest"—that is, a lien or encumbrance over the property under consideration.[1] Under the UCC, a party must secure its security interest by taking two measures. First, it must make sure the security interest "attaches" to the item in question by receiving a specific "grant" of the security interest from its obligor, usually the owner of the property but not always. Second, the secured party must "perfect" its security interest by filing a notice in the proper state office (or offices) so that the security interest becomes a matter of public record that can be located by any party in the future. Once the security interest has attached and been properly "perfected" in all the appropriate filing offices, the priority of the security interest against other security interest holders in the same collateral (if there are any) will thereby be established. Thus, a secured party that receives a grant of a security interest but fails to file the requisite notice, or fails to file the notice in all the different state offices required by the UCC, will generally lose its priority position in the collateral to any other secured parties who have made subsequent but proper filings and to the debtor's bankruptcy trustee, but it will not lose its security interest entirely and can still make its filing at a later date to prevent its priority ranking from falling further. In some states, however, if another secured party knows about the lien of the secured party that failed to "perfect" its interest when the secured party filed to "perfect" its own security interest, that party might lose its fortuitous priority over the other party.[2]

Thus, while title does signify ownership, ownership may be not very valuable, and perhaps even worthless, depending upon what security interests have attached to the property in question. It is not enough for a new owner to require that title be transferred to it, the new owner must also insist that title transfer free and clear of security interests.

In the case of EPC contracts, it is usual to provide that title to any particular portion of the work "pass" from the EPC contractor to the owner as soon as such work is incorporated into the facility, or sometimes even when it ships or arrives at the site in the case of goods, rather than "passing" upon payment by the owner to the EPC contractor. Perhaps this procedure is based on a theory that the sooner the EPC contractor passes title to the owner, the less likely that title will be vulnerable to attachment by the EPC contractor's creditors. The owner, if it is financing its project, must be wary that lenders will require title to transfer as early as possible so they have collateral. Some EPC contractors try to provide that title pass upon substantial completion or provisional acceptance, but a provision like this is not likely to be acceptable to project finance lenders who will want the unfinished work as collateral for their loans to the owner.

More important than when title transfers, however, is that the EPC contractor warrant to the owner that when title does pass to the owner, the title will be free and clear of any liens or claims. For example, in the case of a combustion turbine power plant, suppose the EPC contract obligates the EPC contractor to supply spare turbine blade "buckets" or "baskets" for the combustion turbine blades ("buckets" are racks that hold the different stages of combustion turbine blades). Suppose, further, that instead of purchasing these spare buckets from the combustion turbine manufacturer, the EPC contractor locates a developer who has purchased spare buckets from the same manufacturer but never used them, perhaps because its project was canceled before construction was completed, and the developer has offered to sell these buckets to the EPC contractor at half price. The EPC contractor is eager to pocket the resulting savings because the owner has already signed up to the EPC contractor's lump-sum price. While the EPC contract will contain a warranty requiring the EPC contractor to furnish equipment that is "new and clean" and provisions of the functional specification will usually specify that only spare parts manufactured by the relevant equipment manufacturer may be supplied, the EPC contract is unlikely to prevent the EPC contractor from this type of opportunistic procurement in the aftermarket from a distressed developer. Of course, if the developer has pledged the buckets to its own lenders as collateral for the repayment of the developer's loans and then the developer sells the buckets to the EPC contractor without obtaining a release from its lenders of their security interest in the buckets, the buckets may remain subject to the security interest of the developer's lenders even once they are owned by the EPC contractor and consequently by the owner. This could create two problems for the owner. First, it is conceivable that the developer's lenders could foreclose on their security interest and sell their collateral (the buckets) to satisfy the developer's obligations to them (but the developer's lenders would be required to account to the developer for any difference [excess] between the price the lenders receive in a foreclosure sale and the amount of their loans). Second, the owner will not be able (as its loan agreement will typically require) to grant a security interest in the buckets to the lenders that is first in priority against all other creditors because the developer's lenders already have a security interest in the buckets that will be superior to the security interest of the owner's lenders. On the other hand, if the developer's lenders failed to perfect their interest (perhaps by not filing notices in all jurisdictions required by the UCC), the owner may actually be able to grant a security interest that does have first priority over all other security interests.[3]

Licenses

In some cases (such as patents on equipment and copyrights on computer software), the EPC contractor cannot convey title of these items to the owner. Instead, it must obtain a license for the owner's use of the item in question from the holder of the patent or copyright. In these cases, the owner should make sure that its license to use the product is perpetual, unconditional, irrevocable, transferable to any subsequent owner or operator of the facility and royalty-free. The EPC contractor will usually want to clarify that the license is non-exclusive, which should be of no concern to the owner unless a specific proprietary application has been developed for the owner's facility and the owner does not wish it to fall into the hands of a competitor. Frankly, in most facilities other than perhaps shop drawings of equipment and certain processes such as refining, gasification of fuels, solar power and desalinization, there is probably little technological information that is secret and can generate significant competitive advantages for manufacturers, contractors, owners or operators.

The owner should include provisions in the EPC contract that require the EPC contractor to defend the owner against infringement claims and even require the EPC contractor to replace or substitute infringing items with other items that do not infringe on any party's rights. The owner should make sure that this protection survives the expiration of the warranty period because patents and copyrights remain in effect for long periods and are likely to have longevity far beyond the EPC contractor's warranty period.

Representations and Warranties

In connection with most contracts involving any significant undertaking, the performing party (and generally both parties, even if one party is only paying money for a service) will be asked to make certain statements and then represent to the party for whom it is performing that these statements, in fact, are true. These statements are usually referred to as the "representations and warranties" or the "reps and warrants" or simply the "reps." Their purpose is to assure the party receiving the service or product that certain events have taken place or certain conditions exist.

Usually, the recital of these representations and warranties by the performing party will serve to make the other party comfortable that no investigation (or further investigation) into the other party's state of affairs is necessary because the reciting party has agreed to indemnify the other party for any losses that it incurs in connection with a breach of its representations. For example, an owner building a power plant in a prominent location (such as a college campus) might decide to hire an architect to design an aesthetically pleasing wall to encircle the power plant. In retaining the architect, the owner may ask the architect to represent and warrant that the architect has all the proper state and local licenses and certifications required to work in the jurisdiction where the wall is to be built. Based upon this representation and warranty by the architect, the owner may decide that it does not have to contact all the involved governmental agencies to verify that the architect does, in fact, have all the licenses necessary to render its services.[4] Suppose, further, that after the power plant begins to operate, part of the wall enclosing the plant collapses and seriously injures a student. Suppose, further again, that the attorney retained by the student discovers that the architect was not properly licensed. Consequently, in addition to the student suing the owner for the student's physical and psychological injuries, the student sues the owner for punitive damages for retaining an architect that was not properly licensed. Of course, the student will also probably sue the architect as well, but often the owner will be a "deeper pocket" with more accessible (recoverable) assets than the architect or his firm. After all, the owner has a multi-million dollar power plant sitting right on the campus and probably extensive insurance as well. It is often said that a good strategy for a plaintiff's attorney is to sue everyone that may be implicated in any way in a proceeding and let the judge and defendants worry about whether or not the lawsuit is proper against any particular defendant. In a case such as this, if the owner is required to pay punitive damages to the plaintiff student, the owner will be able to bring suit against the architect for making the misrepresentation to the owner in the contract that the architect was properly licensed. As was seen in Chapter 20, while most contracts specifically exclude payment of special (consequential) damages from a party's liability to the other party, in the case of the student, any payment the owner makes to the student (whether for the student's lost wages, medical bills or emotional harm) will be the general damages of the owner and, therefore, not subject to the exclusion against special (consequential) damages set forth in the EPC contract unless the EPC contract specifically states that these types of damage payments are, in fact, excluded. It will be important for the owner to recover from the architectural firm (to the extent it has assets or has posted a performance bond) because penalties and punitive damages are usually not covered by insurance.

Legal Opinions

In the case of very large or complex contracts such as EPC contracts it is customary to obtain legal opinions about such matters as the EPC contract's enforceability and the legal capacity of the EPC contractor to enter into and perform its obligations under the EPC contract; that is, that no law or court ruling prevents performance of the contract. These opinions are obtained as an added precaution to assure contracting parties that the contract is binding and enforceable as it is written. If it turns out that it actually is not, the aggrieved party may try to proceed against the lawyer who issued the opinion in order to collect the damages that were incurred by reliance on the "incorrect" legal opinion. This is why legal opinions typically note that no one except their addressee may rely upon them.

Notes

1   In legal terms, real estate is known as "real property" and everything else is generally known as "personality" unless incorporated into real property in a manner such that it is not easily separable or removable, in which case it generally becomes a "fixture" and therefore part of the real property in question. Finally, "accounts" and similar items (such as receivables) are generally known as "general intangibles" while statutory constructs such as patents and copyrights are generally known as "intellectual property." The UCC generally applies to personality and "accounts."

2 This result highlights the difference in state legislative philosophies between a legislature that wants to encourage proper filing (hence, a "race to file" state) vs. a legislature that is more concerned about notice and does not want to penalize parties for not filing or improperly filing a notice (or even the state clerk improperly filing a notice) if a second party was in fact aware of the first party's security interest (a "notice" state).

3 Luckily, the problem discussed above usually does not arise in consumer transactions or even commercial transactions involving a merchant selling its inventory in the ordinary course of its business. In such cases buyers acquire title free of any security interests and therefore buyers need not search for filings of security interests against the goods they are buying. For instance, a consumer buying a washing machine from an electronics store or an EPC contractor buying a dump truck from a truck dealer need not concern themselves with the merchant's or dealer's arrangements with the merchant's or dealer's own lenders that are funding their inventory.

4 See, for example, the Title VIII of the New York Education Law, which makes practicing architecture without a license a felony.

Chapter 23

Indemnification and Insurance

In addition to designing a facility that performs poorly, there are many other ways in which an EPC contractor can cause an owner to suffer unexpected losses. Using the example given in Chapter 22 of the student injured by a collapsing wall, the owner will want to be insulated from financial consequences that may arise from such an accident. That is customarily done through an indemnification provision in the EPC contract whereby the EPC contractor agrees to indemnify (reimburse) the owner for any losses and costs the owner incurs in connection with the project. Generally, the EPC contract provisions will limit reimbursement of costs to those that have been reasonably incurred.

While there is a common law right to indemnification in most states in the United States and no contract need exist between two parties for one party to seek indemnification from another party for its losses, it is good practice to include indemnification provisions in contracts between the parties so that the scope of, and procedure for, indemnification will be clear. EPC contractors generally try to limit the purview of indemnification to cover losses that arise from physical injuries to persons or property of "third parties" and that have been caused by the "negligence" or "willful misconduct" of the EPC contractor. The rationale for an EPC contractor's limitation of its indemnification obligations to those losses that result from physical injuries caused by "negligence" is that accidents can happen and, so long as the EPC contractor has taken reasonable steps to avoid an accident, the EPC contractor cannot be held responsible for every vagary. Responsible EPC contractors seldom agree to a more broad form of indemnification than this "negligence" formulation. A broader formulation, such as any injury caused by the EPC contractor whether or not the EPC contractor has acted "negligently," generally will be unacceptable because the EPC contractor's potential exposure for losses probably would not justify its entry into the EPC contract given the inherently dangerous nature of construction projects. As a practical matter, the owner should not dedicate much time and effort to modifying the negligence formulation of the indemnification provision in the owner's favor, especially since, as will be seen later in this chapter, the owner will likely be covered by its own insurance in cases where the EPC contractor does not make a payment under the indemnity. For instance, suppose a serious storm warning is issued. The EPC contractor halts work but fails to tie down a crane, a typical construction practice. The crane is subsequently blown over in the storm and crushes a truck parked on property adjacent to the owner's property and inflicts minor cuts and bruises on the truck's driver, who is dozing in the truck's cab. If the truck owner successfully sues the owner for the cost of removing the immobilized truck, his lost truck rental revenue and replacing the truck with a new truck, the owner will be able to hold the EPC contractor accountable because the EPC contractor was clearly negligent in not tying down the crane. Similarly, if the driver successfully sues the owner for his medical bills and lost wages, the owner should be able to recover these costs from the EPC contractor. Matters will become more complicated if the truck driver sues the owner for his emotional distress because it is likely that the indemnity provisions of the EPC contract will have limited the EPC contractor's liability to physical injuries. Therefore, the EPC contractor is likely to refuse to reimburse the owner for payment of damages for emotional distress.

However, even when the EPC contract is clear on this subject (which it usually is not), this point of contention will not easily be resolved because these are the general and direct damages of the owner, and, it is more likely than not that the owner intended for such damages to be covered by the EPC contractor's indemnity. Again, the court or an arbitrator will look to the intention of the parties if the EPC contract is silent or ambiguous on this issue. And what the parties intended may be a difficult question of fact to resolve. After all, it is the intention of the parties that is dispositive, not the intention of the lawyers who wrote or negotiated the indemnity provisions of the EPC contract. It is the intention of their clients themselves who, perhaps, may not have contemplated the issue at all (or even read the EPC contract's indemnification provision) and who probably cannot advance very plausible arguments regarding their intentions. A more likely scenario, however, is that the truck owner and truck driver will each have his own insurance and make claims against his own insurers, who will pay these claims. Then, the insurers become subrogated to the rights of their beneficiaries (the truck owner and driver), as was discussed in Chapter 21, and these insurers will sue the owner and/or the EPC contractor.

By contrast, if the EPC contractor does in fact secure the crane and it still tips over and crushes the truck and injures the driver, the EPC contractor would probably not be liable to the owner under a negligence formulation of the indemnification provisions of the EPC contract because the EPC contractor has not been negligent. As will be seen later in this chapter, this is why the owner needs insurance. Another interesting question in this case is which party should bear the cost of tying down the crane. That will depend upon the allocation of costs in *force majeure* provisions of the EPC contract.

Sometimes an EPC contractor will attempt further to limit its liability to the owner by attempting to include a provision in the EPC contract that states that the owner waives all rights to recovery against the EPC contractor howsoever such rights may arise under law including, for example, under doctrines of strict liability (which was discussed in Chapter 18) in which the defendant is not entitled to assert any defenses (positive or negative[1]), thus limiting the owner's recovery rights to those arising under the indemnity provisions of the EPC contract. In practice, while the EPC contractor may be successful in including these types of provisions in the EPC contract, as was noted in Chapter 20, these provisions may very well be held unenforceable by a judge because they are against the public policy of allowing a party to escape liability for its own reckless behavior; that is, acting with willful disregard for the consequences of its actions.

Besides providing for the right of indemnification, the owner will want the EPC contract to contain certain other provisions regarding the apportionment of liability between the owner and the EPC contractor.

First, the owner should include a provision in the EPC contract that the EPC contractor will "hold" the owner "harmless" in the case of losses that are covered by the indemnification provisions of the EPC contract (except to the extent of the owner's own negligence, of course). Thus, in the case above, if the truck owner chooses to sue just the EPC contractor and not the owner, the EPC contractor would be barred by this "hold harmless" provision from making a successful reimbursement claim against the owner for the EPC contractor's losses.

Second, the owner will want the EPC contractor to assume the obligation to defend the owner in case any claims arise for which the EPC contractor is liable under the indemnity. Thus, should a potential claim arise, the owner will notify the EPC contractor and give the EPC contractor the opportunity to defend the claim against the owner if the EPC contractor acknowledges that the claim will be covered by the indemnity clause and thereby the EPC contractor can attempt to limit its ultimate liability by reducing, or even eliminating, any award that the owner must pay. The owner should always be careful if it has permitted the EPC contractor to defend a claim to make sure that the EPC contractor agrees in writing at the time the claim arises that the claim is in fact covered by the indemnity because the owner would not want to be in the situation in which the EPC contractor does an unsuccessful job defending the claim and then denies coverage of the owner's loss as not covered under the indemnification provision. In such a case, it might have been better for the owner to defend the claim itself in case it loses the coverage claim it makes against the EPC contractor.

If the EPC contractor chooses not to defend a claim against the owner, the EPC contract will typically provide that the EPC contractor will advance or reimburse the owner for any costs that the owner itself incurs in defending a claim. To protect the EPC contractor from unnecessary payouts, should the owner be tardy in its notification to the EPC contractor that a claim has arisen and thereby somehow prejudice the EPC contractor (perhaps because a court-imposed deadline has passed), the EPC contract will usually release the EPC contractor from its indemnification obligations to the extent that the EPC contractor has been prejudiced by the owner's failure to give prompt notification to the EPC contractor.

If the owner has very significant interests at stake, the owner may not want to turn over defense of the case to the EPC contractor and merely await a result. Therefore, the EPC contract should provide that the EPC contractor must choose legal counsel that is experienced and acceptable to the owner. The owner should also reserve the right to retain its own counsel at the EPC contractor's expense because there can be issues in which the EPC contractor and owner are at odds with each other (such as cross claims in warranty disputes) and also the right to participate in the EPC contractor's defense of claims. Since the owner's reputation may be at risk, as in the case of a hazardous material spill, the owner should also reserve the right to assume the defense of any case which the EPC contractor is not defending faithfully or prosecuting competently. Furthermore, the owner should require the EPC contractor not to settle any civil claim involving non-monetary damages or any criminal claim if such settlement would involve an admission of guilt or plea of "*nolo contendere*"

(Italian for "I do not wish to contend it") on the part of the owner unless the EPC contractor obtains prior consent of the owner before taking any such conciliatory action.

Especially in the case of environmental and patent infringement claims, which can take years to resolve, the obligation to defend the owner or, in the alternative, advance the owner's costs (on at least a periodic basis) can be vital to the owner's financial well-being so that the owner can avoid financing the cost of a litigation or arbitration claim. Unfortunately, seldom will EPC contractors agree to a provision that calls for advancement of the owner's costs. If the EPC contractor will not agree to this provision, the owner may want to compromise on this matter and simply include a provision in the EPC contract that provides that if the EPC contractor is not being diligent in its defense of a claim against the owner, the owner will have the right to settle such claim without jeopardizing the owner's indemnification rights.

Indemnity provisions generally provide coverage for a party's breach of any of its representations, warranties or covenants set forth in the contract and not just in cases of physical damage to persons or property. While a statement of this coverage is helpful, it is probably not necessary in U.S. jurisdictions because one party can always sue the other for a breach of a representation contained in the contract so long as the aggrieved party can demonstrate that it relied upon the breached representation. Including an indemnification provision generally obviates the need to demonstrate to the court that the representation was relied upon, which can be advantageous for the owner as an evidentiary matter. In a breach of covenant, even if such breach is not expressly covered by the indemnity, a party can simply sue for breach of contract. Indemnification provisions are useful, however, because their inclusion will outline a clear process to be followed if a loss or breach occurs.

Quite often, entirely separate indemnification provisions are included in contracts for indemnities related to environmental liability, taxes, intellectual property rights and fines. While it is probably not necessary for these items to be the subject of separate indemnity provisions, sometimes separation of these matters can facilitate easy administration of the indemnity provisions because the parties may want to subject these matters to different procedures, different monetary limitations of liability, different thresholds for liability and different periods of survival for these indemnification rights. The parties may decide that indemnification for environmental claims can be brought any time in the future while deciding that indemnification claims for taxes can be brought only within five years after substantial completion or provisional acceptance. They may also decide that no indemnification claims for taxes can be made until all claims for taxes exceed $25,000 to prevent "nuisance" proceedings. Or they may decide that indemnification claims for a breach of the representations and warranties may not exceed the EPC contract price. Usually, indemnification payments for the EPC contractor's injury to persons or property are not subject to the overall liability limitation of the EPC contractor under the EPC contract (as was discussed in Chapter 20).

The practice of separate indemnities for taxation and environmental matters has become so prevalent that not including specific and separate indemnification provisions could potentially create an implication that the parties did not intend these items to be subject to indemnification rights, a conceivably horrific result. As a precautionary measure, the EPC contract should specifically note that the indemnification provisions cover environmental issues, taxes, fines and violations of intellectual property rights.

While in principle there is no harm in the owner's indemnifying the EPC contractor for any activities to be carried out by the owner, the owner and EPC contractor must recognize that it is the EPC contractor and not the owner that will be in possession of the site and, except in rare instances, carrying out the construction work. Therefore, if an accident occurs, it should be *prima facie* (clear on its face) evidence that the accident was caused by the EPC contractor, or one of its subcontractors, and not the owner. The EPC contract should give the EPC contractor the opportunity to rebut this presumption, but the EPC contract should make clear that it is the EPC contractor's responsibility to bear the burden of proof necessary to overcome this presumption.

Insurance

In addition to indemnification, a party that suffers a loss of any kind will usually look to its insurers for reimbursement rather than the tortfeasor. This is true because the injured party will only have to demonstrate to its insurers that a loss from an insured peril has occurred and generally insurers will then recompense the loss. Once an insurer compensates its beneficiary for losses its beneficiary has suffered, if another party has caused the harm, the insurer will become "subrogated" to the rights of its beneficiary to recover from that party (as was discussed in Chapter 21). In one sense, this is a more efficient result for all parties concerned. The injured party is recompensed and the insurance company, which has the resources and expertise to seek recourse against the party responsible for the damage, is left to pursue the injured party's remedies against the responsible party.

While it is sometimes facetiously said that "insurance companies are companies involved in the business of not paying claims," it is important that this not be the case on infrastructure projects. It is in the best interest of the owner and the EPC contractor to have a good relationship with a project's insurers. A poor relationship can lead to dire financial circumstances if an accident occurs and the insurers do not promptly disburse insurance proceeds so that work can continue without undue delay and disruption. Cases of insureds changing insurers for negligible premium savings can sometimes lead to trouble if the new insurer is not prompt in assessing and reimbursing claims. No matter which party is obtaining the insurance, whether it is the EPC contractor, the owner or a subcontractor, the policy should be obtained from reputable insurers who have experience where the project is located and have a strong financial health rating from an agency that monitors insurance companies, such as A. M. Best's. Sometimes a host government will require that insurance be purchased from local insurers (which obviously serves to promote the local insurance industry). In these cases, the owner should make sure that these local insurers are supported in some manner by creditworthy international insurers (and the owner has privity with these international insurers usually by means of a "cut through clause"). In fact, the owner should make sure that the entity issuing the insurance policy is the entity that is rated by the rating agency and the entity issuing the policy is not a subsidiary of the rated entity. Otherwise, as was discussed in Chapter 14, the owner could be left with a claim against an entity with no assets and no recourse to the insuring entity's ultimate parent.

Builder's Insurance

To cover risks during the construction and testing period, the EPC contractor or the owner will obtain builder's "all risk" insurance (BAR), sometimes called construction "all risk" insurance (CAR). This insurance will cover the work while it is in progress against all risks (such as accidents and *force majeure*) unless a risk is specifically excluded by the policy. The owner must understand what types of peril the insurer is insuring against. Catastrophes that occur suddenly, such as accidents, storms and fire, are generally covered, but instances such as poor workmanship, corrosion, seepage, migration and latent defects are generally not.[2] A latent defect is usually defined as an irregularity not readily observable by a routine inspection. The BAR policy will contain an enumeration of the risks that it excludes from coverage such as war, terrorism, earthquakes and floods. Separate policies must be purchased to cover these risks. Owners must be very careful in their review and negotiation of insurance contracts because, unlike most commercial contracts, their interpretation is generally reserved to the court as a matter of law and not a matter of fact. That distinction will be discussed in the next chapter.[3]

A BAR insurer will compensate the owner or EPC contractor for the full replacement cost (if a replacement cost policy is purchased) of any work lost in an insured accident or peril. If an event of destruction occurs, the insurer, subject to any deductible, will make available proceeds to replace the lost work. Except in rare cases, a BAR insurer will pay if a loss occurs. It is irrelevant whether or not the covered accident or *force majeure* event is attributable to the negligence of any party, even the owner, EPC contractor or a subcontractor. In fact, BAR insurance may even cover instances of sabotage by disgruntled or unscrupulous employees who are seeking retribution, or more work. A good example might be that of an electrician who intentionally miswires a circuit so that it will malfunction and he will be called upon to fix it.

Typically, if the owner is obtaining BAR insurance, the EPC contractor will require that the owner's insurers waive their rights of subrogation to the owner's claim against the EPC contractor so that the EPC contractor is not sued by the owner's insurers. This is logical given that if the EPC contractor had purchased the BAR insurance for itself, the EPC contractor would be making a claim for loss against its own insurance company, which would not turn around and sue the EPC contractor. This requirement of waiver of subrogation rights is, therefore, a consequence of the owner's choosing to procure the BAR insurance instead of requiring the EPC contractor to procure it. The owner should be careful to ensure that the owner is not waiving its right to make claims against the EPC contractor but only the rights of the owner's insurers[4] to make claims against the EPC contractor. After all, if in some rare circumstance such as fraud, the owner's insurers do not cover the loss, the owner should still be able to seek redress from the EPC contractor. (In most cases, special endorsements can be obtained by an insured that provides that the policy will still be effective if another insured has vitiated the policy.)

To whom should the proceeds of BAR insurance be paid? Generally to the party that is bearing the risk of loss and the burden of repairing and replacing lost work—the EPC contractor. Other parties that have what is known as an "insurable interest" (notably, the lenders if the owner has financed the project) in the owner's facility may be named as co-insureds under the BAR policy and, therefore, can claim entitlement to proceeds of the BAR insurance in certain situations.

If the owner is not fastidious in its coordination of these potentially misaligned interests, the owner can unwarily enter into contracts containing conflicting provisions regarding payment of insurance proceeds. In the first instance, an experienced EPC contractor will insist that the EPC contract provide that (unless it has acted with gross negligence or willful misconduct) the EPC contractor will be responsible only to repair and replace lost work to the extent it receives BAR insurance proceeds from the project's BAR insurers. Also, the EPC contractor will not be responsible for advance funds for reconstruction and will await reimbursement by the BAR insurer. Now, in the case that the owner has financed its project, suppose that the owner's lenders have a provision in their loan agreement that provides that, in a catastrophic loss such as an earthquake, the proceeds of the BAR insurance must be paid to them to satisfy their loans if they decide that rebuilding the facility is not feasible, perhaps because it will take too long and market projections for the facility's output or fuel prices have changed considerably since the lenders first lent their money to the owner. Since the owner will probably lose its investment if the lenders choose not to rebuild the facility, the owner should make sure that the loan agreements provide that the lenders' decision not to rebuild the facility in the case of a catastrophe be reasonable and made in consultation with the independent engineer and market consultants, although this requirement may be of little help in practice.

To add a further layer of complication, suppose the owner had been awarded a concession from a government to build the facility and the facility's location or output is strategic. Therefore, the concession agreement required that the owner's facility be pledged as collateral to secure the owner's performance of its obligations to the government or the facility's offtaker.[5] Thus, the government, which has an economic and thereby insurable interest in the facility, will often require under the concession agreement that it be named as an insured on the BAR policy. The government may even require that under all circumstances (catastrophic or not), the proceeds of BAR insurance be used to rebuild the facility, which rebuilding could be in contravention of lender's wishes to use proceeds of the insurance to repay their loans instead. Unfortunately for the owner, if the owner has not highlighted this "proceeds application" issue for lenders before the owner mandates a particular syndicate of lenders, some type of accord will have to be made between the government and the owner's lenders on this issue. If, however, the owner has informed potential lenders of this issue in advance, eager lenders are more likely to concede to the government's "rebuild" condition because the likelihood of a risk materializing is improbable. Paradoxically, a concession agreement will usually provide that an extended event of *force majeure* (typically one year or more) will entitle the government to terminate the concession agreement. In order to address lenders' concern that they will never be repaid if this type of provision is utilized, the concession agreement usually provides (at a minimum) that the government will repay the lender's debt if the government terminates the concession agreement as a result of *force majeure*.

BAR and Expansion Projects

BAR is available with different levels of coverage for different items, such as defects in manufacturing and incidents during startup. As a general matter, the owner's lenders usually insist that the owner or EPC contractor purchase the most comprehensive BAR program available.

In theory, if the EPC contractor has extensive financial resources, no insurance is necessary, and the EPC contractor (and hence, indirectly, the owner) could save the cost of insurance premiums and bear the burden of any cost incurred if a peril arises. This practice is referred to as "self-insuring." Self-insuring is rare except in cases where the cost of insurance has become prohibitively expensive on the basis of the danger involved in the project or where insurance is unavailable (such as for blasting in crowded areas). What is more common is the practice of purchasing insurance with a very high deductible because the premiums will usually be lower. This practice is often referred to as "self retention."

If the owner has financed its project, under the credit agreement, lenders usually require that the owner obtain insurance that meets certain specific guidelines and is otherwise acceptable to them. To address the case in which insurance rates change and certain perils become exorbitantly expensive to insure, the owner should attempt to temper its lenders' requirements by having lenders agree in advance that, notwithstanding any requirements of the credit agreement to the contrary, the owner shall not be required to purchase any insurance coverage not available at commercially reasonable rates. In fact, if the loan agreement requires the owner to obtain insurance and the owner cannot obtain insurance at reasonable rates, the owner will not be able to raise a defense of *force majeure* or impossibility because loan agreements generally do not include *force majeure* provisions and, at common law, this is not the type of obligation that would typically be excused by a defense of *force majeure* or impossibility.[6]

BAR policies are usually purchased for the entire expected duration of the construction period (for example, 36 months) plus some buffer of three or six months and paid for entirely at the outset of the policy. If construction takes longer than expected, the policy holder is usually permitted to purchase two extension periods for an additional premium.

If the owner is expanding or upgrading an existing facility, it is probably wise for the EPC contract to establish which party (the owner or the EPC contractor) will be responsible for damage to the owner's existing facility because BAR insurance typically covers damage to work under construction, not other property such as the existing plant and equipment. The owner's property insurance usually will cover the existing plant, but the owner should verify that this is, in fact, the case under its particular policy. While damage to the owner's existing plant might sound like it would be covered by the EPC contractor's third-party liability insurance, it may very well not be covered because the owner probably will not be considered a "third party" if it owns the existing facility and the expansion facility. The owner usually will be subject to a deductible if damage occurs and it is typical for the owner to make the EPC contractor responsible for this deductible. In fact, erudite EPC contractors often require the owner to waive any claims that the owner may have against the EPC contractor and all of its subcontractors for property damage to the owner's facility that is in excess of this property insurance deductible and even hold the EPC contractor harmless and indemnify the contractor from damages in excess of the owner's deductible. The rationale underlying this request from the EPC contractor is that the EPC contractor is relying upon the owner's property insurance to insure damage to the owner's existing property. Therefore, the EPC contractor will not purchase insurance to cover itself if it causes a loss of this type. Consequently, the EPC contractor should not permit the owner's insurance company to seek indemnification from the EPC contractor if the insurance company reimburses the owner for its loss. Since an insurance company accedes the rights that its insured has through subrogation (as discussed above), if the owner has waived its rights of recovery against the EPC contractor, the insurance company will generally be bound by this waiver. To prevent attempted circumvention of this waiver by insurers, it is customary for potentially liable parties to require insureds to solicit their insurance company's written acknowledgment that the insured has given this waiver.

Marine Cargo Insurance

Marine cargo insurance insures equipment shipped by wet conveyance against loss or damage at all points on its transit route whether or not the equipment happens to be over water at the time of its destruction. The insurer's limitation of liability under a marine cargo policy should never be less than the value of the most expensive single cargo shipment.

When the EPC contractor is procuring the marine cargo insurance, the owner does not really have much to be concerned about. However, when the owner is obtaining the marine cargo insurance, the owner must be vigilant because it is fairly easy to vitiate coverage of the marine cargo policy. The owner must make sure that the EPC contract preserves the owner's rights of indemnification from the EPC contractor if the EPC contractor has been responsible for the marine cargo insurance policy's failure to respond to a loss. For a shipment to be covered by a marine cargo policy, usually the insurer must be given prior notice of any shipment so that the insurer can assess the fitness of the proposed carrier, its vessel and route. If the equipment is very valuable or relatively delicate, the insurer may insist upon the right to witness loading of the cargo onto the vessel to ascertain that the equipment has been properly stowed for voyage.

Consequently, the owner must require the EPC contractor to give the insurer and the owner proper advance notice before a cargo is shipped so that the above protocols can be observed. Although such an inspection sounds elementary and reasonable, an uninformed equipment seller and an EPC contractor that is behind schedule can easily create a problem. If a freighter does leave port without the proper notice having been given to the insurer, the owner should immediately notify the insurer and solicit the insurer's acknowledgment that the cargo will be subject to coverage. If the insurer refuses to issue this confirmation, the owner may be able to make a special arrangement with the shipping line so that the insurer can inspect the cargo at the freighter's next port of call. As a practical matter, the cargo may be buried below other containers in the freighter's hold and the shipper may have little interest in taking valuable time and incurring demurrage costs to unload and reload cargo so that the insurer can inspect the owner's shipment.

Advance Loss of Profits (Delayed Startup)

In connection with BAR and marine cargo insurance, it is possible to obtain insurance for the losses that the owner will incur if commercial operation of its facility is delayed as the result of an event that is covered by either of these policies. In addition to forgoing its profit during the period that completion of its facility is delayed, the owner will also have to meet certain costs such as debt service and, perhaps, even demand charges of its fuel suppliers and transporters. These costs can be insured by purchasing an "advance of loss of profit" endorsement for the BAR policy. In fact, lenders usually require that an advance of loss of profit endorsement be obtained in an amount at least sufficient to cover the owner's out-of-pocket costs (including interest on new loans) but not lost profit because such loss is not the lenders' problem and coverage for lost profits is very expensive. Sometimes EPC contractors will insist that their obligation to pay delay liquidated damages, imposed to compensate the owner for these types of "consequential" costs, be reduced by the proceeds of any insurance that the owner receives for lost profit. Some EPC contractors are even clever enough to insist that the owner's insurers waive their rights of subrogation against the EPC contractor on account of these losses from delays in completion. Both these requests are reasonable and can, in fact, usually be granted by the owner without significantly increasing its insurance costs.

Environmental Clean-Up Insurance

It is common for facilities to be built on "brownfield" sites that have already hosted industrial applications and may already be polluted with hazardous materials. Even a seemingly innocuous industrial use like a glass manufacturing facility could have left behind radioactive materials that were used in the coloring process. If pollution must be cleaned up or sealed off, it is often possible to obtain insurance to cover the owner in case clean-up cost exceeds the cost jointly estimated by the owner and the insurer. Not surprisingly, this insurance is expensive but sometimes its procurement is the only way that a lender will finance an owner's project.

Third-Party Liability Insurance

The most unpredictable losses suffered in construction accidents are those suffered by "third" parties who are not involved in the work. For example, if the EPC contractor is blasting away at a rock formation and the concussion compromises the integrity of the local gas utility's underground gas pipeline, the gas utility's damages could be staggering in terms of repair, environmental damage and customer claims. In fact, owners will often prohibit EPC contractors from blasting. Third-party liability insurance will cover these types of losses. Third-party liability insurance is usually obtained and its premiums are paid on a yearly basis.

Worker's Compensation Insurance

The EPC contractor should be required to obtain worker's compensation insurance as required by law to cover injuries to workers.

Automobile Insurance

The EPC contractor and owner should each be required to purchase their own automobile liability insurance to cover damage to property up to some reasonable limitation on the insurer's liability and to cover personal injury without any limitation on the insurer's liability.

Aircraft Insurance

If the EPC contractor will be chartering aircraft in connection with the work, the EPC contractor should be required to obtain aircraft insurance to cover property, passengers and the crew.

Errors and Omissions/Professional Liability

In most jurisdictions in the United States, professionals such as architects, engineers, accountants and lawyers cannot limit their liability for their professional errors and omissions. As a result, these professionals generally purchase insurance to cover themselves from these risks. Since many of these professionals may have relatively small assets relative to the damage their malfeasance can cause, people contracting for their services often require them to obtain a minimum amount of insurance coverage that can be looked to by them in an attempt to cover losses associated with professional malpractice. Even so, since the professionals tend to work in small firms, insurance coverage of more than $10 to $15 million dollars tends to be rare.

Umbrella Insurance

Depending upon the limitations of liability contained in the various project insurance policies, it may be prudent for the owner to purchase (or cause the EPC contractor to purchase) excess liability insurance to cover circumstances in which an underlying policy limitation is reached by an insurer but losses still remain. In these cases, umbrella coverage will respond once the underlying insurers have reached their payment limitations.

Political Risk Insurance

It is possible for the owner and its lenders to purchase "political risk" insurance, which will insure the owner and lenders against loss of their investment as a result of political risks such as expropriation of assets by a local government or the inconvertibility of local currency into foreign currency. Ordinarily, the owner and lenders must purchase separate political risk insurance policies. Often, as a result of expensive premiums, owners do not purchase political risk coverage for themselves but usually are required to reimburse their lenders for the premiums associated with the lenders' political risk policy. Political risk insurers range from private companies (like AON or Lloyds) to governmental agencies such as the United States Overseas Private Investment Corporation (OPIC) and Nippon Export and Investment Insurance (NEXI) to multilateral agencies such as the Multilateral Investment Guarantee Agency (MIGA).

One noteworthy requirement of most political risk insurance policies is that their coverage can be invalidated if the insured discloses the fact that it is insured against political risks. The insurers' requirement of non-disclosure seems to be premised on the assumption that a host government may be more prone to act in a confiscatory manner if it is aware that an owner's and its lenders' losses will be recompensed by insurance.

As was noted above, a political risk insurer that pays out on a claim will become subrogated to its insured's rights against the confiscating party. Once insurers succeed to these rights, the insurers can pursue their claim (directly against the host government in the case that it breached its agreement with the owner). While the lodging of these claims by private insurers may not be of much concern to a "reformist" or "revolutionary" government, even a rogue government might be more hesitant about provoking the governmental insurance agency of a superpower by acting in a manner that will force one of the superpower's agencies to pay tens or hundreds of millions of dollars in political risk claims. In fact, for countries that have a turbulent history of confiscation, revolution and devaluation, it is probably a wise practice for an owner to involve bilateral or multilateral lending institutions and insurance agencies if it plans to finance its project. As a corollary of this strategy, it also seems advisable for the owner to solicit the participation of local banks and pension funds in its project to the extent they are available. Such a shield standing between the foreign owner and local government may be quite helpful when it comes to deterring or, perhaps, requesting governmental intervention. This is particularly helpful if the owner has acquired a concession in a sector, such as electricity, that the government is accustomed to using as one of its controls (through modulation of the industry's tariffs) to regulate inflation as part of the government's macroeconomic policy. Once an industry that was used to help regulate fiscal policies is privatized, continued use of such a fiscal "tuner" can have drastic consequences on the privatized entities.[7] Fortunately for investors, as will be discussed in Chapter 24, as a result of the seemingly innate mistrust of foreign judicial systems by nationals doing business outside their own country and the "globalization" of the world's economy, useful constructs (primarily arbitration) have developed to alleviate territorial egocentricisms.

Title Insurance

Since an owner typically must secure myriad real estate rights in connection with the development of a project, it is typical for owners and their lenders to purchase (usually separate) title insurance policies to insure that the owner has title (or a valid leasehold) to all the easements, rights of way and real estate that it needs to construct and operate its project. A title insurance policy will insure the real property interests of the owner from claims of other parties. A title insurance company will search the official real estate records to determine whether or not title to the property right in question is clear or whether claims adverse to the owner's interests might exist, perhaps because the property in question was not transferred or granted correctly. A deed might have been improperly executed or filed or transferred in violation of law (such as a wife not being party to a deed in a state that requires joint deeds from husbands and wives) or might have been transferred but remains subject to a lien of a creditor of the original grantor. If an adverse claim exists, the title insurance company typically will not insure against it. As a result, many title policies contain a list of exceptions to the policy's coverage for precisely these types of items. In some sense, title insurance is historical and not prospective like other insurance because insurers research the risks evident from title records and do not insure against these risks. In a sense, title insurance does not insure against claims arising from future events, but rather insures against claims attributable to events that occurred before the policy was purchased.

Operating Period Insurance

Once operations commence, the owner will typically purchase "all risk" property insurance, which will cover damage to the owner's property. Owners and lenders should be cognizant that insurers generally change "all risk" coverage over time. For example, "all risk" policies generally covered terrorism before September 11, 2001 but now terrorism is generally an excluded risk for which separate insurance must be purchased. Since "all risk" property insurance is generally purchased from year to year, if a lender expects an owner to maintain "all risk" property insurance, the lender must understand that the scope of this coverage may change over time. If the lender expects the owner to maintain insurance against "all risks" covered by "all risk" insurance at the time the loan is made, its loan agreement must state this (or at least require the owner to purchase such additional insurance as the lender may, from time to time, request).[8]

Pollution Insurance

The risk of accidental (and sudden) discharges of toxic material by the owner's facility can be insured by a pollution policy.

Business Interruption Insurance

During the operational period, advance loss of profit insurance (as discussed above) will be replaced by "business interruption insurance" if the owner chooses to purchase this type of insurance (which is usually required by lenders). A business interruption policy will provide that, after some deductible period (usually between 15 and 45 days), the insurance company will begin to reimburse the owner for the costs that the owner cannot cover as a result of the facility's inability to generate sales revenue.

Deductibles

When the EPC contractor is providing insurance, the EPC contract should state that the EPC contractor is responsible for policy deductibles. It is not really of great concern to the owner what the amount of these deductibles is because the EPC contractor will bear them. Of course, the owner should want some comfort that the EPC contractor has not set its deductibles so high in order to reduce its premiums that one or two occurrences will wipe out most or all of the EPC contractor's contingency and profit and thereby cause the EPC contractor to lose interest in the owner's project.

If the owner is purchasing the insurance, it is best to agree in the EPC contract that the EPC contractor will be responsible for the deductibles up to a certain agreed-upon level but that the deductibles themselves not be set in the EPC contract. This will allow the owner the flexibility to raise deductibles if it so desires (and thereby reduce the owner's insurance costs) because the owner will bear responsibility for any "gap" between the amount the EPC contractor is required to bear under the EPC contract and the deductible under the owner's insurance policy. The EPC contractor should not be particularly concerned if the owner takes this approach in the case of BAR insurance if the EPC contractor ensures that the EPC contract provides that the EPC contractor will not be obligated to repair or replace lost work until it receives funds from the owner and the insurance company so that the EPC contractor need not advance any costs for reparation.

In the case of third-party liability insurance, however, the EPC contractor might be more concerned because the EPC contractor actually might have to make payment to a third party for damages that the EPC contractor has caused. Thereafter, the EPC contractor will have to seek reimbursement from the insurance company and also the owner if the amount for which the owner is responsible under the EPC contract exceeds the deductible in the owner's policy. This approach may not be the best solution for the EPC contractor, especially if the owner refuses to pay the shortfall because the owner asserts that the EPC contractor has been grossly negligent in causing the loss and, therefore, the EPC contractor is not entitled to reimbursement from the owner. In this case, it would have been better for the EPC contractor to insist upon setting the deductible for the owner's policy in the EPC contract.

Even if the deductibles are set in the EPC contract, the EPC contractor should always reserve the right to approach the owner's insurance companies to lower the level of any deductibles by paying additional premiums to the insurance company. This "buy down" option of the EPC contractor will not impact the owner and will allow the EPC contractor to manage any risks that the EPC contractor believes need to be further minimized.

Effectiveness of Policies

If the EPC contractor is relying upon the owner to obtain any of the project's insurance, the EPC contractor should require that any such insurance is in effect before commencing work at the site. To protect the interests of both parties, the EPC contract should provide that either party can cure the other party's failure to pay any insurance premium and can then "back charge" the other party. To safeguard against lapses in coverage, the EPC contract should require that all policies of insurance obtained by either party in connection with the work include a provision that the insurer will not cancel or terminate its insurance policy without giving prior written notice to all parties named on the policy. All policies should also be "primary" policies and not "excess" or "contributing" policies so that the insureds only need to look to the insurer providing the policy and this insurer cannot delay or disallow any portion of the insurance proceeds because there is another insurer that should be contributing to (or even insuring) the loss in question under a different policy.

Cooperation and Claims

Even if the owner has decided to procure some or all of the project insurance, it is important that the EPC contract provide that the EPC contractor be responsible for filing and prosecuting all insurance claims. This is necessary because it is the EPC contractor that has care, custody and control of the site and the work, and thus the EPC contractor is more likely to be more intimately familiar with the details of any event or accident that occurs at the site. It is also advisable that the owner enlist the EPC contractor's cooperation and assistance in the disclosures that must be made to insurers in order to procure insurance, such as the accident and safety record of the EPC contractor. The EPC contract should also obligate the EPC contractor to comply with any recommendations, such as safety advice, made by insurers.

Subcontractors' Insurance

The above discussion is also applicable to the subcontractors of the EPC contractor, who should have to comply with all requirements to which the EPC contractor is subject so that there are no inconsistencies in any insurance obtained by a subcontractor. Although the owner generally does not have to concern itself with these issues, the owner does not want to be sued by a subcontractor or a subcontractor's insurer. Therefore, the owner should insist on obtaining waivers of the owner's liability from these subcontractors and also waivers of the rights of subrogation of their insurance companies against the owner. From the EPC contractor's perspective, it is important that the owner not be able to use a subcontractor as a "straw man" to bring a claim indirectly against the EPC contractor that the owner is otherwise prohibited from bringing by the terms of the EPC contract. For instance, suppose, as was discussed above, that the owner has released the EPC contractor under the EPC contract from damage caused to the owner's existing facility. Suppose then that the stormwater run-off system of the owner's existing facility becomes clogged with topsoil migrating from the construction site. The owner's insurance company pays the owner's cost of pumping out the owner's drainage system. Then the insurance company sues the civil subcontractor of the EPC contractor for the cost of this damage. The civil subcontractor may respond by impleading (suing) the EPC contractor, claiming that the EPC contractor failed to warn the civil subcontractor of the location of the owner's storm drains. To prevent such legal entanglements, it is advisable for the EPC contractor to make sure that the ambit of the releases and waivers given by the owner and its insurers encompasses the EPC contractor's subcontractors.

Notes

1 A "positive" defense is one in which the defendant admits guilt or liability but claims that it should be absolved for attenuating circumstances. A "negative" defense is one in which the defendant denies guilt or liability altogether.

2 See *City of Burlington v. Hartford Steam Boiler Inspection and Insurance Company*, 190 F. Supp. 2d 663 (D. Vt. 2002), holding that defective wells in the economizer tubes were not a "covered" loss.

3 See *Cresswell v. Pennsylvania National Mutual Casualty Insurance Company*, 820 A.2d 172 (Pa. Super. Ct. 2003), ruling as a matter of law that cracks in the walls of a house were faulty workmanship and that the policy's coverage ceased when the owners "accepted" their home.

4 "Insurers" is the term used here but often construction projects are so large that insurers form a consortium to provide insurance or attempt to attenuate some of their risk exposure by insuring themselves (this is known as re-insurance).

5 Generally, the government and the facility's offtaker will agree to subordinate their liens on the owner's facility to the lien of the owner's lenders in order to enable the owner to obtain financing. As a financial and political matter, the more debt that an owner can borrow to finance its facility (and thereby increase its internal rate of return through the leverage of employing less of its own capital), typically, the more competitive will be the owner's price for the facility's output or use, which should ultimately benefit all the government's citizens. Of course, if the owner is able to achieve a very high ratio of leverage and then the facility runs into operational or technical difficulty, it is more likely that an owner will abandon its project, which could be of severe consequence to the government and its citizens. To guard against this possibility, concession agreements often contain minimum equity investment levels for owners and even change in control provisions that restrict owners from selling their investments in a project.

6 See *Kel Kim Corporation v. Central Markets, Inc.*, 70 N.Y.2d 900 (1987), holding that the inability to obtain liability insurance in a certain amount was not excused as *force majeure* when insurers ceased to issue policies at the liability level required in the lessee's lease.

7 In fact, political risk insurance has developed, in some sense, because it is no longer in vogue to send the militia to protect a foreign investor's interests (as was done by the French government in 1861 when it sent armed forces to Mexico, ostensibly, at least, to collect the "Jecker" loan made to Mexico by a Swiss-French bank). In response to the "diplomatic protection" governments offered their itinerant investors (by assuring that local confiscations would be liberated by military force), certain "indigenous" governing principles began to be propounded by colonial politicians and intellectuals. For instance, the "Drago Doctrine" espoused by Argentine Foreign Minister Luis Drago in a letter to the Argentine Minister in Washington, D.C. in 1902, posited that military force should not be used in the collection of debt from South American governments (which have a history of not paying their national debt: since their independence, Brazil, Argentina, Mexico and Venezuela alone have defaulted 29 times [see *Latin Finance* March 2005]). This proposition was further molded into the so-called "Calvo Clause," which was a provision that began to be inserted into contracts between Latin American countries and foreign investors around the turn of the twentieth century pursuant to which a foreign investor agreed to submit unconditionally to local adjudication of disputes and waive any rights that they might have to seek redress from the courts in their own country for any breach of their contract.

8 See *Omni Berkshire Corp. v. Wells Fargo Bank*, 307 F. Supp. 2d 534 (S.D.N.Y. 2004), holding that "all risk" policies have evolved over time and a lender had failed to protect itself from this possibility.

Chapter 24

Dispute Resolution and Governing Law

While tomes have been written on the resolution of construction disputes and exhaustive coverage of this topic is beyond the scope of this book, a synopsis of some of the issues involved in dispute resolution is helpful for determining how parties can provide for resolution of their issues.

In addition to situations in which one party simply breaches its contractual obligations, disputes can arise as a result of any one of four "congenital" drafting deficiencies in EPC contracts. First, the EPC contract's provisions may be ambiguous and, therefore, the parties do not agree about the intentions underlying the provisions, such as whether or not all subcontractors must individually deliver waivers of their lien rights to the owner when the work is done or it is simply that the EPC contractor must deliver a statement on their behalf. Second, a provision may contemplate the issue in dispute but fail to yield a dispositive result, such as when the parties agree in the EPC contract to replace any index that is no longer published with a reasonable substitute index but fail to agree upon the replacement index. Third, one party may believe a provision is unfair or unenforceable, such as a limitation on liability for recklessness that a party could argue is against public policy. Fourth, the EPC contract may be silent on the issue, such as when a government enacts legislation that provides that all payments under contracts must be made in local currency (as Argentina did concerning loans in 2002 by means of its Public Energy and Exchange System Reform Law).[1]

There are various means available for resolving disputes arising out of EPC contracts and no single solution will be appropriate for all situations. The following sections provide a brief introduction to the various options available to parties when drafting EPC contracts.

Mutual Discussions

Many disputes can be resolved through discussions between parties without having to initiate litigation or arbitration. While not all disputes can be resolved in this manner, it is advisable for the EPC contract to provide that any dispute that has not been resolved promptly by the parties will require very senior (ideally the most senior) officers of each of the parties (or even their ultimate parent entities if the parties are subsidiaries of other entities) to meet in an attempt to settle the matter because they are often eager to avoid negative publicity and undue expense. To ensure that one party does not use these negotiations as a means of prolonging a dispute and to add a sense of urgency to the negotiations, any provision providing for senior-level discussions should also have strict timeframes dictating when the meetings must begin and conclude. If this forum does not produce an agreeable result, more drastic steps should be imposed. While pistols at dawn as a method of dispute resolution has generally gone the way of solid brass railings at power plants, as will be discussed below, many forms of alternate dispute resolution (ADR) have arisen to supplement or replace the traditional routes of litigation and arbitration.

Litigation

Courtrooms may no longer be the repository of stilted manners and officious decorum that they once were, but many states in the United States and also in the United Kingdom are still somewhat merciless in their application of historical legal precedent. In reality, however, strategic and tactical advantages can reverberate from the conservative institutional penchant of the judiciary. Consequently, resolving disputes in front of a judge, but generally not a jury, can be beneficial.[2] A distinguishing advantage of litigation can be the relative predictability of its results. Unless parties are dealing with an issue of so-called "first impression" (which refers to a question of law that has never been adjudged before), a lawyer's application of the law to the facts in a dispute will usually permit the result to be predicted with some accuracy. Of course, results cannot be predicted with certainty but, even if the painstaking appellate process must be exhausted, outcomes and their relative probabilities can be surmised. This relative predictability of outcomes usually has the added advantage of giving the upper hand in settlement discussions to the party who is likely to prevail on the merits of the case in a court. Most courts encourage settlements. Some courts even have programs such as "settlement weeks" during which parties must meet to try to settle their differences out of court and thereby preserve the availability of judicial resources for the citizens that truly need them.

Motions for Dismissal

A party that believes the law is in its favor has the power to make a motion to dismiss a claim on the basis that the claim is not legally permissible, perhaps because the claim lacks an element of the "cause of action" required to plead the claim. Either party also has the power to make a motion to ask the court to rule on a case (a so-called "summary judgment" motion) if it shows that neither the law nor the facts are in dispute (that is, the facts may have been stipulated by the parties). While a party may object to having these motions made against it by the other party, in general, courts are usually not reticent about granting judgment on motions before trial. Although appeals can be made (but will be subjected to a high standard of scrutiny before an appellate court reverses a lower court's decision), the granting of one of these motions can have the practical effect of denying the plaintiff its day in court because the suit will be dismissed or decided (sometimes "with prejudice," which means that the claim cannot be brought again) before the claim is heard in open court—a potential disaster for a party who was relying upon on an in-depth inquisitive proceeding to sway a judge with the totality of the evidence because its case on the law may not have been compelling.

Another tactical opportunity afforded by litigation is that, in addition to the potential for swift disposition of a claim, it is also sometimes possible for a party instead to over-concentrate on the procedural aspects of a case and draw out the adjudication schedule of a suit over a considerable period in an attempt to exhaust its opposition's resources and will.[3]

Considering that perhaps the most likely dispute between an owner and an EPC contractor is the one in which the EPC contractor seeks payment from the owner for work the EPC contractor has done but for which the owner is refusing to pay, from the EPC contractor's perspective neither a swift, but unsympathetic, judicial resolution nor a protracted litigation makes commercial sense and thus EPC contractors tend to prefer mediation and arbitration over litigation.

Choice of Law

In many jurisdictions, parties are generally free to choose which state's or country's law will govern their contract and also choose a state or country to hear their dispute, irrespective of whether or not they have chosen the law of that state or country to govern their contract. EPC contracts for large projects that involve international financing sources are often governed by the laws of England or the State of New York.[4] Both the laws of England and those of the State of New York are relatively well-developed when it comes to construction disputes and therefore allow for a reasonable measure of predictability as to how a dispute will be resolved in court. English law has the further advantage of imposing disclosure obligations on the parties before an action is commenced, allowing for the disclosure of documents that may be adverse to the disclosing party's interest and even criminal penalties for non-disclosure of requested information. The English law policy of the unsuccessful party being responsible for the legal fees of the prevailing party is also said to deter frivolous claims. English law does not recognize the doctrine of punitive damages, which can also be a benefit for EPC contractors whose project performance has been substandard.

For all of the reasons explained above and others, financing parties very often choose New York or English law to govern their own loan agreements and require (as a condition to lending) that New York or English law govern a project's EPC contract as well. The foregoing requirements may very well be imposed by lenders even if the project is located in a civil law jurisdiction.

Judgments

Once a final and non-appealable judgment has been obtained, the prevailing party must seek execution of its judgment if the liable party fails to honor it. Between states in the United States, this execution of a judgment against an obligor is relatively easy unless an *ex parte* ruling has been obtained in which one party failed to appear to defend itself. Contrarily, enforcing a judgment in one country against a citizen or company that is resident in another country can be difficult. Historically, unless a specific treaty is in effect, courts are leery of the dispensation of "justice" in other jurisdictions and have been apt either to decline to recognize or to re-examine court judgments obtained in foreign jurisdictions to ensure that due process was exercised and no xenophobic or political considerations prejudiced the court's verdict.

When to Litigate

Many factors should be considered when deciding upon whether or not to choose litigation as the means of resolving disputes under an EPC contract. By choosing litigation, the parties are agreeing to be bound by the evidentiary and procedural rules of the jurisdiction where the litigation will occur. Thus, for example, by choosing litigation in a U.S. federal court, the parties are agreeing to the applicability of the U.S. Federal Rules of Civil Procedure and the U.S. Rules of Evidence. Likewise, choosing to submit a dispute to the courts of England, France, Argentina or any other country will subject the parties to the particular rules of those countries.

Discovery

Choosing litigation in some countries may subject the parties to extensive discovery (the process of requesting evidence from the opposing party). The United States, for example, has a very liberal process, through which litigants can seek the production of documents and take depositions from witnesses.

Second, litigation provides an opportunity for an appeal on the merits if a party feels that the trial court erred in rendering its decision. Appeals can be useful in that they give a losing party an opportunity to correct an erroneous decision. Appeals, however, can be used as a tactic to delay final resolution of a dispute.

Arbitration

Perhaps in response to the unforgiving impartiality court proceedings' "binary" system of upholding or denying claims without the possibility for compromise, arbitration has developed as a means for parties to submit their disputes to a body empowered to resolve disputes fairly and efficiently.

Like litigation, arbitration is a form of binding dispute resolution. Generally, the parties submit their dispute to one or three arbitrators who are empowered to issue awards that bind the parties to the dispute. Today, in large, complicated, international projects, arbitration is preferred because it allows for a presumptively neutral forum and thus avoids the fear that parties can have about being sued in the other party's jurisdiction. At a minimum, if mediation is included as mandatory in an EPC contract, the parties should impose time limits so as to ensure that the final resolution of the dispute is not unduly delayed.

Most parties believe that arbitration allows a party to have a fair hearing of its case without unnecessary procedural delays and expense. Still, there are procedural issues that can become the subject of debate, such as whether or not the arbitrators have the power to subpoena witnesses to gather evidence. Arbitrators may be less likely than judges to grant motions (such as motions to dismiss or motions for summary judgment) that dispose of a case before it is heard. Presumably, arbitrators are reluctant to dispose of cases early in an arbitration proceeding on the basis that, in the final analysis, the facts will expose any baseless claim and therefore a party should not be denied its opportunity to be heard even if, at the outset, the law does not appear to be in the complainant party's favor. In fact, arbitral rules typically do not even have a mechanism for summary or early dismissal of a case because "non-suiting" an allegedly aggrieved party is not a function of most arbitration rules. Furthermore, one of the grounds for challenging an award is that a party has been denied the right to present its case. This influences tribunals with respect to the evidence they allow and the "due" process afforded parties.

Since arbitrators are not judges, parties to an arbitration could have to deal with what could be called the "King Solomon" proclivity of the arbitral panel—the perhaps human desire to compromise disputes by reaching a fair resolution that apportions liability somewhat equally between the parties so that no party is severely disadvantaged.[5] This need to take into consideration the chance that the arbitrators may be inclined to split the responsibility between the parties has the potential to steer a party that believes that it could prevail on the merits of its case in a court into settlement in an arbitration because the party suspects that its chances of success are less than those that the party has if the matter were being tried in court.

Multi-Party Arbitration Agreements

Occasionally, in the case of projects that involve many parties or that are very complex, a "global" arbitration agreement between all the involved parties may be signed.

"Ad Hoc" vs. "Institutional" Arbitration

Arbitration typically takes one of two forms—institutional or *ad hoc*. Institutional arbitrations are administered by and conducted pursuant to the rules of an arbitral institution. There are a number of such institutions around the world, including the International Chamber of Commerce, the American Arbitration Association, the Stockholm Chamber of Commerce, and the London Court of International Arbitration. These institutions have each promulgated rules designed to govern arbitrations that they administer. They also provide administrative personnel who will oversee various procedural aspects of the arbitration, including the selection of an arbitrator, challenges to an arbitrator, dissemination of communications, and payment of the arbitrators' fees.

Ad hoc arbitrations, by contrast, are not administered by any arbitral institution. The parties to an *ad hoc* arbitration may choose to use arbitration rules promulgated by an institution or they may seek to agree on a bespoke set of rules applicable to their dispute.

Regardless of whether a party chooses an institutional or an *ad hoc* arbitration, it is agreeing to have the arbitrator(s) resolve the dispute in a final and binding manner. As with litigation, there are a number of factors to consider when deciding whether to include an arbitration provision in an EPC contract.

Arbitration provides the parties a significant degree of flexibility in choosing the laws and rules that will govern their proceeding. As discussed above, choosing to litigate in a particular court binds the parties to the procedural and evidentiary rules of that jurisdiction. By contrast, arbitrators are not bound by any national procedural or evidentiary rules. Rather, the parties are free to agree to the procedural rules, evidentiary rules and law that will apply to their arbitration. This flexibility helps ensure that the proceeding is conducted in line with the parties' expectations and helps "level the playing field" when the parties come from different legal backgrounds.

Arbitration is typically confidential and, thus, allows parties to maintain a degree of privacy over their dispute. Litigation, by contrast, is a matter of public record and the proceedings are open to the public unless sealed by a motion made to the courts (which is not liberally granted).

Arbitration is not subject to an appeal of the case on its "merits" and, thus, typically provides a much greater degree of finality than litigation. Unlike in litigation, the parties to an arbitration cannot challenge a final award on the grounds that they believe the arbitrator(s) erred as a matter of law or fact. Rather, the parties may challenge arbitral awards only on grounds that are designed to protect the integrity of the arbitral process generally. For example, parties may challenge an award on the grounds that they were not provided proper notice of the proceeding, that the arbitral tribunal exceeded its authority, or that the award was procured through some form of corruption.

International arbitration awards are significantly easier to enforce than foreign judgments. At the end of either a litigation or an arbitration, the prevailing party is left with a piece of paper setting forth the damages or other relief to which it is entitled. Unless the losing party complies voluntarily, the prevailing party will need to enforce the rights set forth in that in order to obtain its relief. In domestic arbitration and litigation, enforcement is typically not an issue, as courts will recognize domestic arbitral awards and court judgments equally.

In contrast to *ad hoc* arbitrations, parties may choose to submit a dispute to an organization (some governmental, some not-for-profit, and some for-profit) that has been organized to hear and resolve disputes. In 1892 the London Chamber of Arbitration (now the London Court of Arbitration [LCIA]) was set up to arbitrate disputes. The LCIA also promulgates arbitration rules that can be used by parties for *ad hoc* arbitrations. Thirty-one years later, in 1923, the ICC created the International Court of Arbitration in Paris. The ICC also promulgates arbitration rules for parties to use in their own arbitrations. The American Arbitration Association (AAA) followed suit in 1926. Now there are many other organizations, including the China International Economic and Trade Arbitration Commission (CIETAC), which was organized in 1989.

Recognition of Arbitral Awards

One advantage of arbitration is the sanctity of its awards. In international matters, courts are significantly more likely to recognize a foreign arbitral award than a foreign court's judgment. In most countries, the enforcement of foreign arbitral awards is governed by the 1958 Convention on the Recognition and Enforcement of Foreign Arbitral Awards (also known as the New York Convention) or the Inter-American Convention on International Commercial Arbitration, a similar treaty adopted by the Organization of American States in 1975 (known as the "Panama Convention"). These treaties require that, subject to limited exceptions, signatory states enforce foreign arbitral awards issued in another signatory state in the same manner that they would enforce a judgment from their own courts. There are over 140 parties to the New York Convention and 19 parties to the Panama Convention.

The cost of arbitration must also be considered. Unlike litigation, where the costs of the court are paid by the state, the parties to an arbitration are responsible for paying the arbitrator's fees and the costs of any arbitral institution chosen to

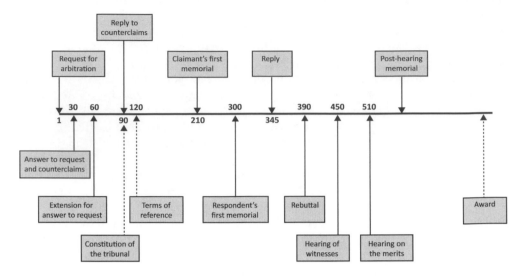

Figure 24.1 Timeline for a typical ICC arbitration.

administer the proceeding. The costs will vary depending on the size and complexity of the matter and the extent of testimony. Discovery can be extremely costly and time-consuming, thereby extending the time necessary to resolve a dispute. Many (if not most) common law jurisdictions similarly permit various types of discovery. Civil law countries typically do not have the type of discovery permitted in the United States but generally do have their own rules regarding the production of evidence and documents. The institutional rules generally do not provide for motions to dismiss or summary judgment applications. Thus, unless the parties are able to reach a settlement, it is likely that an arbitration will be resolved only after a full hearing on the "merits" of the case. Figure 24.1 shows a timeline for a typical arbitration administered by the International Chamber of Commerce.

In international projects, these factors can cause some concern, as at least one party likely will not be from the chosen country and, thus, may be unfamiliar with the applicable discovery rules.

Despite many xenophobic concerns, there can be distinct advantages to litigation. First, unlike arbitration, litigation typically provides various options for achieving a summary resolution of a dispute. In the United States (and many jurisdictions), for example, a party may move to dismiss a case at an early stage on various grounds, including lack of jurisdiction or the failure to state a legally cognizable claim. In addition, parties may also seek summary judgment of a case. As mentioned above, summary judgment will be granted when there are no factual issues in dispute and the court will only have to rule on a party's legal entitlement. Both of these procedures permit courts to resolve disputes without having to go through a full hearing of evidence.

Although arbitration does not generally involve the stringent requirements of litigation, it remains a structured proceeding employing rules either promulgated by the parties themselves or adopted by the parties from a pre-existing model form of rules. For instance, in 1976 the United Nations Commission on International Trade Law (UNCITRAL) published a code of rules that can be used by parties to guide them in arbitrating matters concerning international trade. Whether the parties rely on some published codification of rules or use their own self-devised rules, of key importance will be the number, appointment, independence and qualification of the arbitrators included. Most arbitrations have either one arbitrator or three. In the case that there will be three arbitrators, usually each party will choose one arbitrator and the two arbitrators will then attempt to agree upon the third. If they cannot agree, an arbitral organization (such as the ICC) selected by the parties could be designated to choose the third arbitrator. The parties may wish to require that the third arbitrator be a lawyer. The parties may also want to provide that the arbitral panel must issue a reasoned written award in connection with its resolution of the matter. The utility—and danger—of arbitration is precisely that the parties can be the architects of the rules of the forum in which they will argue their cases.

ICSID Arbitration

In 1966, the member states of the International Bank for Reconstruction and Development (the World Bank) signed the Convention on the Settlement of Investment Disputes between States and Nationals of Other States, creating the International Centre for Settlement of Investment Disputes (ICSID) in Washington, D.C. This treaty provides an arbitral mechanism for foreign investors to seek relief against a host government (so long as the host government has signed the treaty and agreed to arbitration in the contract underlying the dispute). While it is still a formidable task to seek recourse against a government, ICSID has helped foster the notion of non-discriminatory treatment of foreign investors and has been instrumental in eliciting inherently non-liquid investments in infrastructure projects in many developing nations (as opposed to liquid investments such as stocks and bonds, which can usually be readily sold if a problem arises, albeit usually at a loss). For the EPC contractor or owner that is contracting directly with a government, the failure to include an ICSID arbitration clause in its contract could very well prevent the owner or EPC contractor from obtaining financing from international banks, who take comfort in relying upon the availability of ICSID arbitration.

Mediation and Conciliation

In addition to arbitration and litigation whereby the parties submit requests for relief to a neutral body for resolution, sometimes the parties choose to seek advice in handling their dispute from an independent third party.

If the parties to an EPC contract are unable to resolve their disputes through mutual discussions, or if a dispute persists following a decision by an independent expert or dispute resolution board (each as discussed below), they may attempt to resolve their dispute through mediation. Thus, the parties might appoint an independent facilitator, mediator, conciliator or technical expert to review their dispute and propose a non-binding solution for the parties to consider before they proceed to arbitration or litigation.

Mediation is a formal dispute resolution process by which the parties engage a third party to assist them in reaching an amicable resolution. Unlike litigation and arbitration, in which courts and arbitral tribunals are empowered to issue binding decisions, a mediator has no authority to resolve a dispute or issue any binding decisions. The mediator's role is to facilitate discussions between the parties and, in doing so, it can perform a variety of functions. In some cases, the mediator will conduct "shuttle diplomacy," in which she moves between the parties to convey their positions and offers but does not interject her own views as to the relative strengths and merits of the parties' positions. In other cases, a mediator will advise the parties as to the strengths and weaknesses of their cases and how their positions would likely be received by a judge or arbitral tribunal. Ultimately, the mediator can play as active or passive a role as the parties desire.

In fact, organizations such as UNCITRAL and the ICC promulgate rules for conciliation in addition to their rules for arbitration. A number of courts and arbitral institutions encourage, or at least provide, a framework for mediating disputes. For example, in the context of litigation, there are a number of state and federal jurisdictions that either require mediation or authorize courts to refer civil actions to mediation.

In the arbitral context, the American Arbitration Association and the International Chamber of Commerce have created rules for conciliation in addition to their rules for arbitration. Specifically, the International Chamber of Commerce handles mediation through its ICC International Centre for ADR. In January of 2014, the Centre implemented the ICC Rules for Mediation, which are specifically designed for mediations conducted under the Centre's auspices. Similarly, the International Centre for Dispute Resolution (the international arm of the American Arbitration Association), the London Court of International Arbitration and the Stockholm Chamber of Commerce each employ a set of Mediation Rules that allows parties to seek mediation prior to, or as a component of, arbitration.

While this mediation approach has little risk owing to its non-binding nature and can help a party determine the relative strength of its case based upon the reaction or suggestions of the mediator, the process can have the undesired effect of delaying relief. Depending upon what type of relief is requested and the importance of the claim (that is, a disagreement over whether or not the punch list has been completed vs. whether or not the EPC contractor is behind schedule), mediation can be beneficial. All things considered, however, it is probably not advantageous for the owner to agree in the EPC contract that mediation or conciliation is mandatory and that any dispute between the parties will be submitted for mediation or conciliation before such dispute can proceed to arbitration or litigation. At a minimum, if mediation is included as mandatory in an EPC contract, the parties should impose time limits so as to ensure that the final resolution of the dispute is not unduly delayed.

Independent Experts

Often parties to an EPC contract will agree upon a list of "independent technical experts" (usually updated from time to time in lengthy projects) who can be called upon to either suggest resolutions or finally resolve disagreements that may arise regarding technical subject matter of the EPC contract. This practice is generally helpful, but many disputes encompass more than just technical issues and involve contractual interpretation. Therefore, this procedure can be inappropriate for many controversies, and the owner should impose appropriate limitations on what disputes can be resolved by the technical expert. Nonetheless, courts and arbitral panels generally will uphold the technical expert's pronouncements if the parties intend to be bound by them. The decisions of the technical expert typically are final and binding unless one or both of the parties formally challenges the decision. At that stage, the decision will be subjected to review by a court or arbitral tribunal, depending on which forum is provided for in the EPC contract. While the decisions of the independent expert are accorded some degree of deference, the parties are able to argue the merits of their cases to the court or arbitral tribunal and the review is not intended to be merely a rubber stamp of the expert's decision. The practice of using independent experts is common in contracts for the construction of commercial buildings wherein the owner designated its own architect as the final arbiter of design specifications, and thus the architect is given the authority to determine whether the contractor is acting in accordance with the owner's specification.[6]

Dispute Resolution Boards

Dispute resolution boards (also referred to as dispute adjudication boards or DRBs) are an alternative to the use of an independent expert. As a preventative measure (usually for very large projects), the parties may set up a formal or informal board of experts to review progress of the work from time to time. This board will usually be well-positioned to settle any disputes that may arise. Decisions of a DRB typically are binding on the parties unless one or both of them formally challenge the decision. At that point, the decision may be submitted for review by a court or arbitral tribunal in accordance with the procedures specified in the contractual dispute resolution clause.

While a DRB is very helpful for technical engineering and construction problems, it is really not suited to deal with problems caused by macro-economic or political forces. One distinct advantage of dispute resolution boards and independent experts is that, unless the parties have agreed otherwise, they can issue "advisory opinions." Essentially, one or both parties can seek advice or guidance as to what a provision or specification means before a dispute actually arises between the parties. This right of consultation can be useful in avoiding disputes entirely because one party may realize that the technical expert or dispute resolution board is not likely to concur with its analysis of the situation. Courts will not provide the same advisory service. Courts provide the public service of resolving controversies, not issuing advice. For a court to "hear" a case, there must be a matter in dispute, not a matter that could lead to a dispute. Issuing advisory opinions is generally believed to be a waste of public resources. On the other hand, while courts do not issue advisory opinions, they can use their equitable powers to issue declaratory judgments to make a declaration resolving an issue actually in dispute so that the parties will have direction in how to continue performance of their obligations in accordance with the contractual provision under debate.[7] Thus, unlike courts that require a controversy before they will entertain proceedings, these less formal mechanisms of dispute resolution, such as appointing experts and dispute resolutions boards, can be helpful in resolving ambiguities in contractual provisions or the functional specifications before a controversy arises.

Pendency of Disputes

It is critical that an EPC contract provide that the EPC contractor not be able to suspend or slow its performance during the pendency of a dispute. Even with a good dispute resolution program in place, disputes generally take time and strain parties' resources. The EPC contract should ensure that the existence of a dispute will not divert the EPC contractor's attention from the prime directive—completing the project on time.

Disputes with Governments, Agencies and Regulated Utilities

Owners should take note that in contracting with governmental or regulated entities that are subject to supervision of a regulator, contracting standards and/or bidding rules, these entities may be very reluctant to settle any dispute (or even make a payment to settle a matter that they do not dispute) unless an award or judgment has been entered against them. A public official or employee can only be subject to criticism, scrutiny or dismissal for making a payment without an order of an arbitrator or judge to do so. With this in mind, in contracting with these entities the owner might want to consider a provision for accelerated dispute resolution or "fast track" or similar arbitration such as Judicial Arbitration and Mediation Services (JAMS).

Language

The EPC contract should specify which language will govern the interpretation of the EPC contract (and arbitration or mediation of such are provided for), notwithstanding the fact that negotiations may have been conducted in other languages or that translations of the EPC contract may have been prepared for the parties' (or if the project has been financed, their lenders') convenience. The parties should also specify whether or not these "unofficial" translations may be used in helping interpret any ambiguous provisions of the "official text" of the EPC contract.

Costs

Under all of the above non-judicial proceedings, parties are generally free to allocate the costs of the proceedings as they choose, and the EPC contract should address how costs will be shared or shifted between the parties.

Notes

1 Most "change in law" provisions contained in EPC contracts are limited in scope to changes in environmental, labor and tax law and typically would not offer protection to the EPC contractor in the case in which the EPC contractor's costs rise because it is suddenly being paid in local currency although the EPC contractor may still be incurring significant costs

(for major equipment, for example) in hard currencies (which tend to appreciate in value against inflationary local currencies).

2 Absent special circumstances, the EPC contract should provide for a waiver by each party of its right to trial by jury. Generally, there is little advantage for either side to have a complicated construction dispute resolved by a jury of the uninitiated.

3 In fact, Napoleon III abolished the courts of his uncle and it is said that no one cared because most of the parties involved in the pending cases were either dead or doting at the time of the courts' abolition.

4 Generally, in the United States, claims are heard in state courts and not federal courts unless the claim concerns a question of federal law (so-called "subject matter" jurisdiction) or the parties involved in the lawsuit are from different states (so-called "diversity" jurisdiction), in which event a case can be "removed" from a state court to a federal court.

5 In fact, psychology experiments have shown that all parties (even the winning party) are usually more comfortable if a resolution of a conflict is fair rather than overly burdensome for one party to the detriment of the other party.

6 See *Savin Bros., Inc. v. State*, 405 N.Y.S.2d 516 (N.Y. App. Div. 1978), holding that a contractor could not make a claim for extra work where the contractor had agreed to be bound by the decisions of the owner's engineer (unless the contractor could demonstrate that the owner's engineer had acted in bad faith).

7 See *Milford Power Co., LLC v. Alstrom Power, Inc.*, 822 A.2d 196 (Conn. 2003), holding that even though the EPC contractor had given notice of the occurrence of a *force majeure* event, since an actual claim for an equitable adjustment had not yet been made, no justiciable question had yet been presented to the court. Therefore, the case was dismissed.

Chapter 25

Interested Third Parties

Assignment of the EPC Contract

As a general matter, parties have the right to delegate and assign their duties under a contract to another party unless the contract expressly forbids any such assignment or delegation. In fact, in the United States, the right to assign payments cannot be prohibited. This notion is in line with the U.S. principle of "freedom of contract," which provides that parties are free to enter into contracts between themselves on any terms so long as these terms are not contrary to public policy. However, U.S. courts will not "find" terms in a contract if the parties have not included these terms in the contract (on the theory that the parties were free when they entered into the contract to include any terms that they so desired). If the parties chose not to include a particular term (consciously or in oversight), it is not the court's responsibility to negotiate or reform contracts, simply to interpret and enforce them.

Assignment of a contract or delegation of one party's duties to another will be permitted unless the parties have chosen to prohibit such an assignment or delegation. The initial contracting party, of course, will remain liable for the performance that has been delegated unless its counterparty or the contract expressly excuses it. One exception to this policy of unrestricted "assignability" arises with respect to contracts for personal services and guaranties. In these cases, the purpose of the contract or guaranty is performance by the contracting party or guarantor itself and not one of its assignees. First, the assignee might not have the credit standing that the guarantor does. Second, even if the assignee has the same creditworthiness as the guarantor does, as a commercial matter the beneficiary may have decided that it has enough credit "exposure" to the assignee because it has made other loans and accepted other guaranties from the assignee and, therefore, it may be imprudent for the beneficiary to expose itself to the risk inherent in concentrating too much of its credit exposure in one entity.

Whether or not EPC contracts are contracts for personal services or goods may vary from jurisdiction to jurisdiction, but the owner will not have to worry about this issue if the owner restricts the EPC contractor from assigning any or all of its rights under the EPC contract. Sometimes the assignment restrictions contained in a contract will note that either party "may not assign any of its duties or rights under the contract." Some lawyers believe that this language can create an implication that, while discrete duties cannot be assigned, the entire contract can, in fact, be assigned. While this interpretation is dubious, the necessity for a court or arbitrator to decipher this cryptic text can be avoided by the simple recitation that "neither this contract nor any duties of a party hereunder may be assigned." The EPC contractor may be allowed to "subcontract" some or all of its work if the EPC contractor complies with the conditions that the owner has made applicable to subcontracting (as discussed in Chapter 7), but the EPC contract should be unequivocal on the point that any subcontracting of the EPC contractor's work will not relieve the EPC contractor of any of its responsibilities under the EPC contract.

The EPC contractor has much less at stake in the case of an assignment by the owner. As was discussed in Chapter 10, the owner's primary obligation is to pay the EPC contract price. Nevertheless, the EPC contractor will still often want some "veto" over the owner's assignment rights in order to protect the EPC contractor's interest in cases of assignments that could adversely affect the EPC contractor. Generally, the EPC contract will require that the EPC contractor give its consent to any proposed assignment by the owner unless there is a reasonable justification for its withholding consent.

It is highly unlikely that the owner will ever need to avail itself of the assignment provisions of the EPC contract (unless a taxation or internal organizational issue has arisen and as a result it has become advantageous for the owner to assign the EPC contract to one of the owner's subsidiaries or affiliates).[1]

If, however, the owner is planning on selling the project before construction is complete, the owner should pay careful attention to assignment provisions of any parent guaranty that the parent of the owner has given to the EPC contractor. As was discussed in Chapter 2, if the owner desires to sell its project, this can usually be most efficiently and quickly achieved by the owner's selling the SPV that has been created to carry out the project. However, if the EPC contractor has obtained a guaranty of this SPV's performance under the EPC contract from the owner's parent, this guaranty will remain in effect in spite of the sale of the SPV to another party. Thus, if the owner is contemplating the sale of the SPV to another party, it behooves the owner to place a provision in the guaranty that allows the owner or the owner's parent to assign the guaranty to another party in the case that the owner sells the SPV (because the owner would not want to remain liable on the guaranty in such case). While, understandably, the EPC contractor would not want this assignment to occur without the EPC contractor's prior consent (because it may not be comfortable with the reputation or credit standing of the proposed guarantor), it is usually a reasonable compromise between the EPC contractor and the owner to prescribe so-called "safe harbor" provisions that will provide that if the proposed assignee of the guaranty satisfies a pre-agreed-upon set of criteria (usually financial stability and technical qualifications), the assignment can be made without the EPC contractor's consent. If this approach fails, either because the EPC contractor will not agree to include "safe harbor" provisions or because the proposed assignee does not meet some or all of the "safe harbor" criteria, the owner still has options. One option is that the owner could continue with the sale of the SPV and its parent remain liable under the guaranty but the parent guarantor require the purchaser of the SPV to indemnify the parent guarantor for any amounts that the parent must pay out under the guaranty to the EPC contractor. This is not the best resolution for the owner or its parent because its parent will be taking the credit risk that the buyer of the SPV will not honor the indemnity if the guaranty is ever called upon. There could be many reasons for this refusal to honor the indemnity, including the buyer's insolvency, the buyer's dissatisfaction with its new investment in the project, or the buyer's trying to set off the indemnification amount that the buyer owes to the parent against the amount that the buyer claims it is owed by the owner under the SPV purchase agreement (perhaps, as compensation for a breach of a representation relating to the project such as a representation that no cost overruns are anticipated). Thus, rather than simply collecting payment under the indemnity, the owner might have to pursue a lawsuit. To mitigate and compensate the owner for taking these risks, the owner might want to consider charging the buyer a fee to keep its parent's guaranty to the EPC contractor in place and also ask the buyer to post collateral or a letter of credit to secure the buyer's indemnity obligations in respect of the owner's guaranty to the EPC contractor.[2]

Fortunately for the owner, it is rare that the EPC contractor will attempt to restrict the owner's ability to sell the owner's SPV. Even if the EPC contractor has restricted the sale of the SPV, the owner may often be able to circumvent these restrictions by effecting a merger with a potential acquiror if the EPC contractor has not crafted "change in control" provisions in the EPC contract designed to prevent a sale of the SPV to another party through other means such as a merger (or a sale of another entity that is "higher up" in the chain of ownership). However, in the case of a merger, even though the SPV's successor will now be responsible for the obligations under the EPC contract, the parent will still remain liable under the guaranty. In legal nomenclature, a "successor" usually refers to a party that has acceded to another party's interest by operation of law without any contractual or consensual action involved (as would be the case under a bankruptcy law, a merger statute or an intestate statute). An "assignee" or "assign" typically refers to a party that has acceded to another party's interest by way of a consensual agreement.

Publicity, Confidentiality and Proprietary Information

Unfavorable media coverage and community opposition have the potential to disrupt or topple just about any proposed project. The proposed "Westway" tunnel project in New York City is a good example. Even some of its strongest proponents have acknowledged that had the proposal been given another name such as the "Tunnel," it might have avoided the strong community opposition that it encountered and that perhaps prevented "Westway" from becoming one of the largest highway projects ever built. A similar project in Boston, which burrowed a tunnel for the Fitzgerald Expressway (often referred to as the "Big Dig"), did not succumb to similar environmental opposition and seems to have received the lion's share of the Federal Highway Trust Funds that could have been used to build Westway.[3]

Careful coordination of the owner's media program is required to prevent miscommunication to the press. The EPC contract must restrict the EPC contractor from issuing public announcements or press releases without the owner's prior written consent. The owner should also prohibit the EPC contractor from including the owner's name or a description of the owner's facility in the EPC contractor's advertising materials without the owner's reviewing such materials in advance so the contractor does not unknowingly disclose sensitive information about the owner or its counterparties. In fact, while there is very little about most infrastructure businesses that is confidential, typically EPC contracts contain confidentiality provisions that forbid either party from disclosing information received from the other (except to the extent necessary to perform its obligations under the EPC contract, to develop and solicit investor and lender participation in the project, as required to be disclosed by law). Expressions such as "on a need-to-know basis" should always be avoided because they are difficult to construe and open to differing opinions on what "need" really means.

Indiscreet Conduct

While perhaps all countries have laws against bribery and graft, the standards and enforcement of these laws may vary.[4] Indeed, some countries have enacted laws that restrict the behavior of their citizens in foreign jurisdictions. For example, in the United States, the Foreign Corrupt Practices Act of 1977 (FCPA) forbids a U.S. entity from even offering anything of value to a "foreign official"—and "foreign official" is very broadly defined.[5] Thus, while an EPC contractor obviously does not have to be told to comply with local laws regarding corruption, it is often a good idea for the owner to alert the EPC contractor to the laws of any other jurisdiction that the EPC contractor should observe in addition to local law.

While personal gain is legally unacceptable in governmental situations, public gain is usually encouraged. Communities or municipalities that are willing to host projects often are the recipients of parks, schools, hospitals, shelters and other amenities. Whether the owner is supporting these ancillary projects out of its own charitable motivations or its zeal to complete its project with the least public resistance, it is satisfying to see that Adam Smith's "invisible hand" is also helping to build social facilities in addition to private infrastructure. Of course, owners must be certain that their "donations" are permitted under the laws of their home jurisdiction and should seek legal or governmental guidance before embarking on any "public works" program. For instance, in the United States, its Department of Justice will issue opinions on whether or not a "donation" will violate the FCPA.

Notes

1 In legal taxonomy, a "subsidiary" is an entity that is owned, directly or indirectly, in full or in part, by another entity. "Affiliates" are entities that are under the common control of another entity. In some cases, statutes provide a definition of "affiliate" to be used in connection with the application of the statute itself. For instance, an "affiliate" for purposes of the federal securities laws is any entity in which the subject entity controls or owns more than 5 percent of voting power thereof.
2 Unfortunately for the owner, there are financial implications involved in keeping its parent's guaranty to the EPC contractor outstanding. GAAP and IFRS will usually require that the guaranty be noted as a contingent liability in a footnote to the financial statements of its parent and, therefore, analysts, creditors and customers of the parent may view this potential liability negatively irrespective of whether or not the buyer has agreed to "backstop" this indemnity with collateral.
3 Lopate, Phillip, *Waterfront* 106 (Crown Publishers, 2003).
4 In fact, Transparency International publishes "Bribe Payers" Index and Corruption Perceptions Index for people to use as a reference.
5 15 U.S.C. 78dd-1. (a) **Prohibition**. It shall be unlawful for any domestic concern, other than an issuer which is subject to section 30A of the Securities Exchange Act of 1934 [15 USCS § 78dd-1], or for any officer, director, employee, or agent of such domestic concern or any

stockholder thereof acting on behalf of such domestic concern, to make use of the mails or any means or instrumentality of interstate commerce corruptly in furtherance of an offer, payment, promise to pay, or authorization of the payment of any money, or offer, gift, promise to give, or authorization of the giving of anything of value to—

(1)   any foreign official for purposes of—

(A)   (i) influencing any act or decision of such foreign official in his official capacity, or (ii) inducing such foreign official to do or omit to do any act in violation of the lawful duty of such official, or

(B)   inducing such foreign official to use his influence with a foreign government or instrumentality thereof to affect or influence any act or decision of such government or instrumentality, in order to assist such domestic concern in obtaining or retaining business for or with, or directing business to, any person;

(2)   any foreign political party or official thereof or any candidate for foreign political office for purposes of—

(A)   (i) influencing any act or decision of such party, official, or candidate in its or his official capacity, or (ii) inducing such party, official, or candidate to do or omit to do an act in violation of the lawful duty of such party, official, or candidate,

(B)   inducing such party, official, or candidate to use its or his influence with a foreign government or instrumentality thereof to affect or influence any act or decision of such government or instrumentality, in order to assist such domestic concern in obtaining or retaining business for or with, or directing business to, any person; or

(3)   any person, while knowing that all or a portion of such money or thing of value will be offered, given, or promised, directly or indirectly, to any foreign official, to any foreign political party or official thereof, or to any candidate for foreign political office, for purposes of—

(A)   (i) influencing any act or decision of such foreign official, political party, party official, or candidate in his or its official capacity, or (ii) inducing such foreign official, political party, party official, or candidate to do or omit to do any act in violation of the lawful duty of such foreign official, political party, party official, or candidate, or

(B)   inducing such foreign official, political party, party official, or candidate to use his or its influence with a foreign government or instrumentality thereof to affect or influence any act or decision of such government or instrumentality, in order to assist such domestic concern in obtaining or retaining business for or with, or directing business to, any person.

(b)   **Exception for routine governmental action**. Subsection (a) shall not apply to any facilitating or expediting payment to a foreign official, political party, or party official the purpose of which is to expedite or to secure the performance of a routine governmental action by a foreign official, political party, or party official.

(c)   **Affirmative defenses**. It shall be an affirmative defense to actions under subsection (a) that—

(1)   the payment, gift, offer, or promise of anything of value that was made, was lawful under the written laws and regulations of the foreign official's, political party's, party official's, or candidate's country; or

(2)   the payment, gift, offer, or promise of anything of value that was made, was a reasonable and bona fide expenditure, such as travel and lodging expenses, incurred by or on behalf of a foreign official, party, party official, or candidate and was directly related to—

(A)   the promotion, demonstration, or explanation of products or services; or

(B)   the execution or performance of a contract with a foreign government or agency thereof.

Afterword

This book is intended to help prepare professionals for the monumental task of developing an infrastructure project on time and on budget—a feat that is very satisfying economically and psychologically but rarely achieved. Often, there are good reasons why projects are late or over budget. Unfortunately, these good reasons often become the subject of debate, arbitration, litigation and bankruptcy proceedings.

Construction projects are inherently dangerous. Overruns will cost some people their jobs. Poor performance will cost some their reputations. Accidents will cost some their lives. The goal of this book is to expose the risks involved in developing infrastructure facilities so that these risks can be estimated and allocated among project participants. The biggest risk is the risk that has not been allocated, whether or not it was contemplated. If it has been allocated, it can often be easily and efficiently dealt with should it arise. The structural integrity of any contract is threatened by any risk that has not been explicitly addressed. If there is only one lesson to be learned from this book, it is that known risks should be examined and allocated—not be ignored and entombed.

An honest and enlightened evaluation of the risks inherent in the development of infrastructure facilities can make the difference between a physical and financial disaster area and a showcase of engineering fortitude and monetary reward. We all know the story of Androcles who took a thorn out of the lion's paw and then they became the best of friends. Of course, that is a fairy tale with no didactic value about assessing risk. The truth is that the lion ate him.

Table of Case Citations

Index/Glossary

Printed and bound by CPI Group (UK) Ltd, Croydon, CR0 4YY

21/10/2024

01777180-0001